ACCESS TO THE AIRWAVES
My Fight For Free Radio

ACCESS TO THE AIRWAVES
My Fight For Free Radio

by Allan H. Weiner
As told to Anita Louise McCormick

Loompanics Unlimited
Port Townsend, Washington

This book is sold for informational purposes only. Neither the author nor the publisher will be held accountable for the use or misuse of the information contained in this book.

Access To The Airwaves

My Fight For Free Radio
© 1997 by Allan H. Weiner and Anita Louise McCormick

Published by:
Loompanics Unlimited
PO Box 1197
Port Townsend, WA 98368
Loompanics Unlimited is a division of Loompanics Enterprises, Inc.

Cover art by Shaun Hayes-Holgate

ISBN 1-55950-163-4
Library of Congress Card Catalog 96-79424

Contents

Preamble

"Far better is it to dare mighty things, to win glorious triumphs, even though checkered by failure, than to rank with those poor spirits who neither enjoy much nor suffer much because they live in the gray twilight that knows neither victory or defeat."

— President Theodore Roosevelt

Introduction
by Andrew Yoder

"Talk hard!"

This line was made famous when uttered by Christian Slater in *Pump Up the Volume*. Well-known to viewers because it is one of Christian Slater's first major films, the movie is an inspirational look at pirate radio — a frustrated teenager trying to liberate the airwaves with his own personal views and musical tastes.

Pump Up the Volume was the first taste of pirate radio for most of those who saw it. It started off with the electronics-whiz/teenager, Harry, playing "Wienerschnitzel" by The Descendents and venting his frustrations via the radio transmitter that he bought "to talk with his friends back East." The programs eventually metamorphized into an inspiration to his entire high school, which culminated in the firings of corrupt school officials and the dramatic FCC arrest of Harry (Slater).

OK, that was just Hollywood. How about real life? Instead of running a fictitious story line, maybe the producers of *Pump Up the Volume* should have gone with *The Allan Weiner Story*?

I can't speak for everyone, but my interest in pirate media began in 1977. I saw an NBC newscast that included one brief segment about the pirate TV station Lucky 7, which broadcasted for a weekend from Syracuse University in New York. I thought that the prank was incredibly cool, but being a 10-year-old, I could do little to watch the broadcasts or contact the perpetrators.

The next major turnaround was in 1980, when one of my best friends got a World War II shortwave receiver — a Hallicrafters

SX-28A. We were amazed at hearing radio stations crackle through the static late at night. While looking for information on shortwave radio in 1981, I found something that clinched my interest for good. *S9*, a then-dying CB-radio magazine, branched out into alternative media, including pirate radio. I found a copy (the only copy that I've ever seen on a newsstand). My quest became: 1. Get a shortwave radio, 2. Listen to the pirates!

By 1983, I had achieved the first goal, but #2 was a little tougher. I listened for months and didn't hear anything. Finally, in June 1983, I heard my first pirate, WRAM (a station that I think has received more attention from me mentioning that it was the first station that I ever heard than from its actual broadcast activities). My second pirate was Radio Clandestine, a powerful and professional send-up of international, clandestine, and rock radio. And the third was KPRC, Pirate Radio Central, on 1616 kHz from New York City.

Unfortunately, my receiver only covered the shortwave amateur radio bands, and KPRC (and a few other stations from New York) only operated just above the AM broadcast band. So, I listened to them on an AM/FM/cassette player, with the whip antenna cranked up. I was living in Western Pennsylvania, so to hear a little pirate station from New York City on just a portable stereo was awesome. KPRC was also the first station that I telephoned, nervously, late at night, talking quietly so that my parents couldn't hear that I was making a long-distance call.

Pirate Joe's call-in show on KPRC was always very interesting. The call-in topics ranged from U.S. foreign policy in El Salvador to "what pirates were on the air this past weekend." At the time, I occasionally grew tired of the politics on KPRC, but I eventually learned to appreciate the programming and viewpoints (even though I often disagreed). In 1984, KPRC disappeared without a trace — no farewell show, no goodbyes, and no glimpse of future plans.

Later that year, a station called KPF-941 popped on the air, claiming to be a legal broadcaster. It sounded like a pirate, though, as if the guys from KPRC were just having fun above the AM dial.

I even received a letter from J.P. Ferraro (along with Allan Weiner, one of the main announcers) insisting that KPF-941 was not a pirate. Apparently, he was right. The station did have a license, but the FCC overruled it and closed the station down anyway.

About this time, I went to college and kind of forgot about shortwave radio. I picked a good time to let the tubes cool, because the pirate radio scene was dead — barely any pirates bothered to grace the airwaves on AM or shortwave (some FM pirates broadcasted from New York City, but because of the limited range, they had only local audiences). During this time in the mid- to late-1980s, pirate radio came to a near standstill. The pirate radio clubs in North America either folded or barely scraped by.

In the summer of 1987, I was working as a construction "gopher," between years of college. Every day, I got up at 6 a.m. and weakly poured a bowl of cereal. Then, I slowly ate it and numbly watched the NBC pre-*Today Show* news on a little black-and-white TV in my parents' kitchen. On the morning of July 27, 1987, I was stunned to see overhead views of a North American pirate radio ship, known as Radio New York International (RNI). Strangely enough, the NBC *Nightly News* triggered my interest in pirate radio and the NBC morning news resuscitated my interest in listening.

After hearing some quick sound bites from the news clips and catching the RNI show that evening myself on 1620 kHz, I realized that RNI was a combination of the KPRC/KPF-941 crew and personnel from the AM/FM pirate, WHOT, plus a few other pirate radio veterans. Amazing. Before long, stories about Radio New York International were splattered across magazines and the front pages of newspapers around the world. Of course, the saga ended with the crew in handcuffs, watched by guards with machine guns, no less! In just a few days, RNI had grabbed the attention of the world and been martyred. As a final show of respect for the fallen pirates, *Rolling Stone* named RNI the best radio station of 1987.

Not only did RNI pique my interest in pirate radio, but it jump-started the entire pirate radio scene. Some of the first pirates on the air after RNI were WENJ, WCPR, and WKND — all of which used 1620 and 6240 kHz, which were old RNI frequencies. Pirate radio was back.

In the summer of 1988, I began writing *Pirate Radio Stations: Tuning in to Underground Broadcasts* — the first book about pirate radio in North America. Of course, one of my first moves was to look up the infamous Allan Weiner so that I could ask for information. It was, after all, Weiner who was behind KPRC, KPF-941, RNI, and some of the best-known early North American pirates. I stumbled through, leaving a message on his answering machine, but I never received a response. Little did I know that he had more pressing matters — RNI was about to stake another claim on the airwaves.

The station returned on October 15, 1988 on 1620 kHz. Just like in 1987, RNI remained on the air for only a few days, then disappeared after the government issued a restraining order against them. This time there was no media fanfare, no guest appearances on MTV, and no awards. Was pirate radio getting too frumpy to attract attention?

No way! *Pump Up the Volume* hit the theaters soon afterwards. The writers were no doubt inspired by the popularity of RNI with the pop culture and counter-culture. Then top-rated Los Angeles FM rock station KQLZ changed formats to a looser style and called itself "Pirate Radio," complete with a Jolly Roger logo. I just happened to be waiting outside of a mall near Philadelphia, Pennsylvania when a black Winnebago, dubbed "the pirate ship," pulled in. It was a local radio station's official mobile studio, arriving for a remote broadcast. Even if RNI lost tens of thousands of dollars, these FCC-licensed stations were discovering that the pirate image was financially profitable.

In the early 1990s, RNI went legal, without all of the excitement, bells, and whistles of the 1987 broadcasts. This time, the station was on with airtime purchased from a large commercial shortwave station in Nashville, Tennessee. For a while, RNI performed predictably. Taking advantage of the powerful worldwide signals,

Pirate Joe kept the old KPRC sound going with call-ins and liberal politics. Johnny Lightning and friends were airing oldies rock, Steve Cole broadcasted radio news, and special guests, such as Allan Weiner, filled in with a variety of programming. But the station (and its programming) gradually disintegrated, and it was evident that RNI was facing internal problems. After several years, RNI went silent.

By the mid-1990s, pirate radio had exploded again in North America. Unfortunately, the same could be said for Radio New York International. Although the station was inactive, the old broadcast ship, the *M/V Sarah*, received worldwide attention again when, instead of transmitting to the world, it was destroyed with high explosives at the end of the movie *Blown Away*.

While I was starting work on an updated edition of *Pirate Radio*, I noted that RNI was apparently gone for good. But, although RNI was off the air, Allan Weiner pulled through again — before I had finished the book. This time, a new ship, the *Fury 5* was being set up to air RNI programs and those from other organizations. Given RNI's past failures, many radio hobbyists sat back and waited to see what would happen. Would the FCC raid the station? Would it sink? Would it be raided by terrorists?

On Christmas morning, 1993, I listened to a terribly weak station broadcasting an old RNI program, but who was airing it? Although I still don't know the answer, the broadcast did show the future of the station. On January 19, 1994, the FCC raided the ship and cut out all of the transmitting gear because it said that the holiday broadcasts were emanating from that equipment. Impossible — the transmitters weren't yet functional and weren't capable of emitting the type of signal that the Christmas Day pirate transmitted.

At this writing, Allan Weiner is working hard again, this time on Lightwave Mission Broadcasting and a new radio ship, the *Electra*. Will this ship station be quickly closed down, like the others? Will it regain the international attention that peaked during the RNI 1987 broadcasts? It's anyone's guess. Regardless, you can bet that Allan Weiner will continue to fight for his access to the airwaves.

Should the producers of *Pump Up the Volume* have gone with *The Allan Weiner Story?* Nah, no one would have believed that it all really happened.

— Andrew Yoder

Foreword

Dear Reader,

You may remember me from Radio New York International — the offshore radio station I designed and built back in 1987 to broadcast free-form rock and roll and messages of peace, love, and understanding to all who wished to listen. You may also remember how the United States government rewarded me for providing this free radio service — they boarded my ship, arrested me, and put me in handcuffs, while federal agents went down below and destroyed the radio station it took me a year and a half to build.

It was another chapter of my life — I build radio stations, and the FCC always seems to find a way to destroy them or rip them away from me.

I know, who cares about my personal life? But everyone has a story to tell. And so do I. My story is one of a lifelong fascination with the art and science of communicating by means of radiant energy transmitted through space in the form of radio waves. Ideas, inspirations, and goals touch everyone's life in some form. And the art of radio has touched mine.

I believe that radio should be fun. And radio should be free, like the rays of the sun. I know — radio frequencies are a limited resource. Without some kind of government control, the radio bands would be chaos. But I still don't agree with the way the government handles it. Instead of licensing radio frequencies in a way that best serves the public interest, they now sell them off to the highest bidder.

I am no traditional broadcaster. I hold radio communications as sacred as the printed word. What follows is my tale, entrenched in the world of the controlled electron with government regulations, personal egos, and songs of the mariner.

Enjoy these words as much as I have enjoyed broadcasting so many over the air, from land and from sea.

— Allan H. Weiner

Chapter One
In The Beginning

I was born on June 12, 1953, at 11:50 AM at the Yonkers Professional Hospital in Yonkers, NY. Yonkers is about 15 miles north of New York City proper and 10 miles north of Manhattan. Back then, Yonkers was known as "The City of Gracious Living." Now, of course, it's one of those cities that's going down the great slide of urban decay. In fact, I think it's already reached the bottom. But when I was a child, Yonkers was a nice place to live, and the apartment our family rented at 166 Hilltop Acres was a pretty comfortable place to be in.

I did not have a boring childhood. I was always very inquisitive. My father is an attorney, and that's probably part of it. His mind has always been sharp and inquisitive, and I guess part of it rubbed off on me. I was always investigating one thing or another and trying to figure out what made it work.

As a young child, the high point of my day was crawling under the living room table with my father's tools and pretending I could fix something. My father kept his tools in an old metal can. I used to drag it out and play with the pliers and screwdrivers and whatever else he kept in there. I never liked the fake tools they made for children — I always wanted real tools that you could do real things with.

My interest in trying to figure out what made things work sometimes caused some very unusual things to happen. One day when my mother was shopping at Gimbels' Department Store, it took her too long to decide which pocketbook she wanted to buy. I got bored and wandered off in search of something more fascinating. Then without any warning, all the lights in the store went out. My mother immediately realized I was gone and proceeded to get hysterical.

Before long, Gimbels' security guards found me. It seems that I had wandered into the back room, found the master lighting switch, pulled it, and turned off all the power to the building. Now why I did this, I do not know. My parents said that the security guards were pretty unhappy about it. But I was in seventh heaven.

That experience of shutting the lights off at Gimbels' was one of the early turning points in my life. After that, I knew for sure that, "Yes, I want to control electrons!"

Of course, I did the things that nearly all kids do when I was growing up. On warm summer days, I liked to go out in the yard to dig holes and play with the garden hose. I didn't soak people down or do anything like that — I just liked to play with it. I really liked the nozzles and thought they were fascinating devices. When the weather was too inclement to play outside, I filled up the sink with water and played in it. I've always loved the water, whether it be a river, a pond, or an ocean. I guess that translated into my later interest in ships and sailboats.

And many wonderful afternoons of my childhood were spent sitting on the floor of our kitchen, tearing all the spoons, pots, and pans out of the pantry, and talking to my mother. I used to ask her, "What's this?" and "Why is that?" about every subject imaginable. And every time I looked at something, I wanted to investigate it, explore it, tear it apart, and figure it out. That's one of the nice things about being human, I guess. It's our nature to be inquisitive about things. We see something, and we want to know what makes it work.

As a little kid, I always wanted to crawl behind our television set and see what was inside of it — even though my father warned me that if I did, all these outrageously horrible things would happen to me. Like I'd blow my head up or be attacked by monsters or something like that. I grew up believing in TV monsters because my dad knew that was the only way to prevent me from crawling back behind the TV to take a look. I remember asking him, "Why is there a television picture?" And he'd say "Well, there are little people in there." I never bought that. The first time I broke open a picture tube, I was really disappointed. There was nothing inside but a lot of glass and something called an electron gun.

My mom and dad inspired me to pursue a broadcasting career more than anyone else by far. My mother's name is Anastasia Borowitz Weiner. Mom was a real sweetheart. I loved her very much. She was a beautiful woman and had very nice features. She was 5-foot-8, had light brown hair, and her eyes were hazel like mine.

My parents got married in 1945 during the war. Mother was a Catholic of Russian decent. My father is Jewish, and his ancestors came from Austria. Mom started off as a career lady, got married, and had my sister Barbara and me. She did what a mom would do back in the '50s. She took care of us kids, made us nutritious meals during the day, and propped us up in front of the television set to watch all of those great black and white cartoons they aired back in the '50s. Then, when we were old enough, she went back to work as a secretary. My mother and I always had a good relationship. Up until her death in 1970, we could talk about anything.

My father's name is Samuel Weiner. He's a lawyer who handles corporate, divorce, and real estate cases. Dad is about 6 feet tall and weighs around 170 pounds. He's a warm person and a very loving father. When I was 6, 7, and 8 years old, he took me to the electronic stores in Westchester County on the weekends — stores like Allied and Lafayette Radio. I can still remember what a thrill it was to go there and search for parts to build everything from crystal sets to Morse code oscillators.

I must have been about 6 or 7 years old when I built my first crystal radio. It was one of those kits you could get in a box for $1.99. I put it together and made it work, but it was kind of boring because there wasn't all that much to it. Just a few coils, a variable capacitor, a crystal detector, and a set of headphones.

When I was growing up in the 1950s, real electronics were in vogue. And everyone was talking about all the new consumer electronic products that were out on the market. Back then, everything had to be hi-fi — just like it all has to be digital today. My father had a Magnavox hi-fi set that had two big woofers and a tweeter, and I used to play with it all the time.

Whenever my parents took me to visit any of our relatives, I always checked out the television — if there was one, or the radio — which there always was. As soon as I saw a piece of electronic equipment, I lunged for it. I must have been the ultimate nerd.

My Uncle Irving had an old Emerson portable tube set that I played with every time I had the opportunity. I was at his apartment eating dinner one evening when he told me that the Emerson had died. Uncle Irving noticed the way I had been gazing at the old set and said, "Allan, would you like to have it?"

I said, "Sure, I'd love it!" And I was in seventh heaven once again.

My father walked over to see what I was getting myself into. You know how parents are. I was about six years old at the time and he wanted to make sure it was something I was ready for and that I wouldn't get electrocuted from it. My parents always worried that I would electrocute myself. (And yes, I've shocked myself many, many times. But I'm still alive.)

I could hardly wait to get that Emerson radio home. We couldn't eat dinner fast enough. I wanted more than anything to get out of that apartment, take that radio home, open it up, and explore it. Up to that point, I had only been allowed to touch and tune radios. And of course, I was allowed to listen to them. But I had never before been permitted to delve into their inner workings.

So the moment we got home, I took the radio out of the paper bag and immediately opened up the back. Then I looked in at all the wonders that I was later to figure out — all the coils, capacitors, resistors, and other parts that make the magic of radio a reality.

I must have spent a week fooling around with that radio. I got it to work, but it didn't last for long because I promptly got my father's can of tools out of the closet and took it apart. I not only took it apart — I dissected it. I took every single part out of that radio — every single screw, bolt, lock washer, resistor, capacitor — everything! My favorite part was the speaker. My father gave me a guide book so I could identify the components. So once I finished taking all the parts out, I proceeded to identify them, label them, and figure out what they did.

From that point on, I was really hooked. There's something about electronics, broadcasting, and the communications media. It doesn't matter how old you get, once you get interested in it, that's it — you're hooked for life. After that, nothing was safe. If it was a radio, television, or anything electronic, I'd try to get my hands on it, take it apart, figure out how it worked, and I was usually able to fix it.

My father used to get a big kick out of taking me over to friends' houses and having me explain how their radios and television sets worked. Sometimes, I also got to manipulate their hi-fi systems. Most people were pretty hesitant to let a 6-year-old fool around with their electronic equipment. But for me at least, it was fun.

My childhood is one of mostly good memories. I consider myself very fortunate in this weird world of constant child abuse. My upbringing was solid, my parents were fine people, and I had a happy childhood. My parents provided very well for my sister and me. We had what we needed and my parents were very encouraging with my extracurricular activities.

My parents got along, but not the best. At times they got into a lot of arguments. Occasionally, their arguments got physical and things flew around the room. But somehow my interest in exploring what makes radios and televisions work helped to shield me from the effects of their arguments. Maybe it was my ability to focus on whatever I was investigating — what makes this happen and what makes that happen — that protected my mind from my parents' arguments and fights.

Unfortunately, my sister Barbara, who is five years older than I am, had a rougher time dealing with it all. Even as a kid, she was always in some kind of turmoil or distress. She had a lot of stomach pain that was probably caused by all the stress and tension in the house. My mother was always a good mom to me. But I think to my sister, she was a little harsh.

In 1959, just before I started school, my father bought a colonial brick house a few blocks away from the apartment complex we lived in. It had three bedrooms, three baths, and a fantastic basement, which I eventually took over with all my electronic projects.

Home in Yonkers, New York, where I grew up.

My kindergarten and elementary school years were spent at Yonkers PS 21, a school that was built in the 1920s. It had beautiful murals and art deco stuff. Structurally, it was a fortress. I remember my kindergarten experiences more than any other early grade in school. I had a wonderful time in kindergarten. We did all kinds of great things — finger painting, drawing, coloring, craft projects. It was fun.

But first grade was entirely different. That's when you have to get into serious stuff like reading, writing, and arithmetic. But the thing that really bothered me in school is that you have no personal rights. I soon realized that the whole educational system is a setup. You have no rights and it is basically a dictatorship from first grade until you get out of high school.

When I entered first grade, I learned that everything was the wonderful world of NO. No, you couldn't do this. No, you couldn't do that. You couldn't leave when you wanted, you couldn't talk when you wanted, you couldn't even go to the bathroom when you wanted. And that was a big, big revelation for me. In first grade, the whole world is realigned. You are NOT an individual any more — you are just somebody who is told what to do. Now mind you, there were rules at home, but the rules were administered by my parents and they were meant to protect me. But the rules at school made absolutely no sense at all.

While both of my parents were very supportive about my electronic interests, they were always concerned that it would interfere with my schoolwork. I hated school. Of course, there were certain parts of the elementary school experience that were all right. But most of the time, it was miserable dealing with the dictatorship and being cooped up with all those kids who were screaming and yelling and treating each other lousy.

In first, second, and third grade, my favorite activity was show-and-tell. Everything I brought in for show-and-tell had to have something to do with electronics, whether it was a tape recorder, a microphone, a vacuum tube, or some of the stuff I built. I brought in amplifiers and PA systems, which were a big deal to me at the time because you could hear your voice over the speaker. Sometimes the other kids in class really liked the stuff I brought in, but sometimes they didn't. And show-and-tell was always a flop when even the teacher didn't understand what I was talking about.

I'd hear kids say things like, "He's crazy — he plays with radios." The only thing that helped me get through school was the fact that I was taller than almost everyone else in the class, which worked to my advantage because it kept most of the bullies away.

When I was growing up, my father was always into gadgets. And he always bought the newest type of radio. This early exposure to the world of technology undoubtedly contributed a lot to my interest in radio and things electronic. I remember when he brought a shortwave radio home and, boy, did I have a blast with that.

I remember well the day my father turned me on to shortwave radio. We were sitting in his bedroom and he had his new Emerson 8-transistor shortwave radio on his lap with the whip antenna pulled out. He tuned across the shortwave bands — and all of these wondrous noises came through the speaker. It was great! We must have spent hours tuning over the shortwave dial and picking up dozens of foreign radio stations broadcasting in languages that neither of us could understand.

My father also tuned in some of the early broadcasts of W2XMN, which was the first high-powered FM station. It was built by Edwin Armstrong in 1937 to experiment with FM transmissions in the 40-50 MHz range.[1] He heard broadcasts from the Alpine Radio Tower, the tower on the Empire State Building, and various other locations. He listened to FM mainly because he was into classical music — and that's what the FM stations of the

---

[1] Edwin Armstrong built W2XMN after he was kicked out of his Empire State Building office by RCA. He conducted his first experimental FM broadcasts from there to show the world that FM (frequency modulation) was superior in coverage, noise figure, and fidelity. I grew up listening to the station that Major Edwin Armstrong built. In the late '50s it was still on the air as W2XMN, an experimental FM station. It later switched its frequency to the standard FM band.

time were playing. Even as I child, I recognized that FM (frequency modulation) was superior in audio quality to AM (amplitude modulation).

I don't know exactly when I got interested in the transmission side, but radio as a magical electronic medium took me right from the start. The excitement of it all was endless. Every chance I had, I would go the library and read whatever I could find. There was a lot I couldn't really understand at that age, but I just kept reading until I understood it.

The late '50s and early '60s were the years of the hi-fi boom. Kit building was the craze. Everyone was building their own radios, televisions, and hi-fi systems from kits. There was Knight-Kit, there was Eico, and of course there was Heathkit. One day my father, who was always into what's new and what's going on electronically, decided that he wanted to build a hi-fi system. So he ordered a Heathkit power amplifier with an AM/FM tuner and a turntable to go with it.

When I found out about it, I was really excited. I was only six or seven years old at the time, but I can still see him taking it out of the box and setting up all of the cardboard cartons and cups that held the parts, hardware, and so on.

The instruction books at the time were very good. The only problem was that my dad is color-blind, and kit building is a matter of connecting the green wire from point A to point B and looking for color-coded resistors and capacitors. If you're color-blind, you're going to have a problem. So whether he wanted me there or not, I was immediately enlisted as the color interpreter. I was by his side every moment he was assembling it.

My father had me strip the wires to the prescribed length. And every once in a while, he let me use the hot soldering iron to solder some of the terminals. He claims that I was eventually able to help him build it by just looking at the schematic, bypassing the instruction manual.

I sat with him for hours, watching him assemble this stuff. Eventually, I was doing half of the mechanical and electrical work on it. After about a week, we got the thing together. And believe it or not, it worked the first time! Later on when I was in high school, I built lots of kits — shortwave receivers, test equipment, audio oscillators, and so on. It was a good way to save money. You get high-quality components and it's lots of fun to put them together.

That hi-fi set got lots of use in our house. My mother used to listen to all the "beautiful music" stations and my father liked to listen to the classical music stations in the Yonkers area. I remember many an afternoon walking into our house and hearing Mantovani and 100 Strings coming over the radio. When I built my first pirate station in 1968, all of the music I played on the air was classical music and beautiful music because that was all I had available.

By the time I was eight years old, I was in the acquisition mode. I would do anything to get hold of more and more radios to play with and make work. I loved vacuum tubes and I found the circuitry and electronics of radios totally fascinating. And then, to my father's chagrin, I started the long process of junk collecting.

The '50s and '60s were the junk collector's dream years. In the late '50s, it's incredible what people threw away. All kinds of wonderful consoles — radios that would probably be worth thousands and thousands of dollars today — were just lying there on the street. Especially when Yonkers had spring cleanup time and people put out radios, televisions, refrigerators, and whatever. I walked down the street and there they'd be — the RCA consoles, the Atwater Kents, the old 1940s portable radios — just sitting there waiting to be picked up.

I wanted like anything to grab this stuff, but my father would have none of it. I remember one day in particular my father and I took a walk up some street in our neighborhood and every house had something that, to a kid of 7 or 8 years old, was fantastic. Old radios, old television sets — some of the stuff was bigger than I was, but I wanted it ALL. I wanted to take it home, dissect it, figure out what made it work and get it to operate if I could. And so began my long career of junk picking.

Junk collecting is still a big part of my life. There is always an economic angle to everything you do in radio. It takes money — it doesn't matter if it's a ham station, a CB station, or whatever. And at an early age I realized that you could find a lot of this stuff very cheap or for free by scrounging around the neighborhood for old radios that could be stripped of their tubes, capacitors, resistors, parts and what have you. Then you could use the parts to build all kinds of other stuff. When I was in the 4th, 5th, and 6th grade, my friends and I sometimes took a Saturday or weekday afternoon just to walk around the neighborhood and look for old radios.

In those years, people were throwing out really good stuff. Now it's all disposable plastic crap. But back then, some really nice sets were going on the scrapheap. In fact, that's how I got my first real shortwave radio set when I was about 12 years old. I was coming home from school one day and I happened to notice this beautiful console radio sitting out next to the street. I think it was a Spartan. It was a huge thing with nice big AM and shortwave dials on the front. I immediately stopped and went over to look at it. Then a lady came out of the house and said, "Hey, if you want it, you can have it."

And I said to myself, "This radio is coming home with me!" There wasn't anyone around to help, so I had to carry and drag that set for a quarter of a mile or so. It took me a while to get it home because it was heavy and I didn't want to damage the case.

I took it down to the basement, cleaned it up, took the chassis out, and replaced the line cord with one I took off of an old lamp. I plugged it in and lo and behold, it worked! I was astounded. I kept saying to myself, "How could somebody throw something away that actually works?" It had a few crunchy controls, but nothing else seemed to be wrong with it. I hooked it up to an antenna and could hear all kinds of shortwave stations coming in from overseas.

But then my father came home. He saw it and immediately got upset because I had dragged in somebody else's garbage. That was his big thing — "You're filling up my house with everyone else's trash."

And I said, "This isn't garbage — this stuff is GREAT!"

Before long, I was finding radios everywhere. At one time, I must have had fifty or a hundred of them scattered around the basement. It got so bad that my father refused to go down there. Even so, I have to attribute the fact that I got interested in the electronic side of radio to my father.

After collecting all of this equipment, I realized, "I don't want to just listen anymore, I want to transmit!" I had an old 1926 Zenith in the basement and asked myself, "I wonder if I hooked a microphone to where the speaker is and changed a few things, would it be possible to turn the radio into a transmitter and broadcast?"

I thought about it for about 15 or 20 minutes, then realized that without major circuit modifications, it was a ridiculous concept. But the idea of transmitting had occurred to me, and putting my own radio station on the air just seemed to be the natural course of action.

For Christmas or Hanukkah, my father bought me a pair of Lafayette CB walkie-talkies. My friends and I used to play with them all the time. They only had a range of about 100 feet or so. They were a lot of fun, but I never really considered that broadcasting — it was just point-to-point, one-person-to-another communication. That kind of thing didn't really interest me. That's why I never had much of an inclination to learn Morse code or become a ham radio operator.

I wanted to broadcast to people — meaning the general public — on a band that a regular radio could pick up. Now that, to me, was exciting.

When I was in the 5th and 6th grades, my friends and I talked about broadcasting a lot. One day, Kenny Sofer and I saved up some money and went to Lafayette Radio to buy our first AM broadcast-band transmitter — the Lafayette Broadcaster. It consisted of a small metal cabinet, two tubes, and a very small radio oscillator that put out less that 1/10th of a watt. It operated under Part 15 of the FCC rules, which meant that you didn't have to get a license to use it. It came complete with its own antenna, which was a 5-foot piece of wire that hung out the back.

The Lafayette Broadcaster was one of the first real kits that I built on my own without my dad's help. It was really a simple circuit, and it only took me one night to build it. I meticulously followed the instructions and soldered every part very carefully.

After I put it together, I brought it upstairs to my parents' bedroom and set it up on the window sill. I threw the antenna out the window. Then I hooked up the crystal microphone, switched it on, and turned on my father's Emerson portable AM radio. The first thing I heard was the feedback of the microphone howling because the radio was sitting too close to the transmitter. But I was enthralled. I thought, "My heavens, this is something! Now I can actually talk over the radio — not a CB set — but a REAL radio!"

I called my friend Kenny, who lived just across the street from me, and told him to turn on his radio. He picked it up, but the signal was extremely weak. We put a second Lafayette Broadcaster at Kenny's house up in his bedroom on the second floor, and we tried every way imaginable to extend the range. We hooked the Lafayette Broadcaster up to the water pipes, telephone line, old car antennas — ANYTHING to get the signal out a little farther. But no matter what we did, we could only get about half a block out of it, and that was all.

Kenny, myself, and another friend, Paul Rosenberg, formed a small radio station we called KPSR. We went on the air at night and played phonograph records. We sat the microphone in front of the phonograph speaker. Then we spent hours playing radio station with the horrible old records our parents owned. Considering the fact that the signal only went about 400 or 500 feet, I don't think we really had any real outside listeners — but at least we could broadcast to each other. And that was fun.

Looking back on it, it's amazing that they sold Lafayette Broadcasters to the general public, because they had an all-metal case and if you plugged it in the wrong way, you could really get shocked with it. One day, we had it setting on top of a cast-iron radiator. We plugged it in, turned it on, and BOOM! All the lights went out when the case shorted to ground.

But by then, the broadcasting bug had really bitten me. I think I got so desperate one day that I took one of my CB walkie-talkies, pushed the "talk" switch on, and taped it in that position. Then I mounted it in front of the phonograph speaker grill and put on a record. Then I walked around the neighborhood with my other CB walkie-talkie and listened to the music. I told myself that it was almost like a real radio station — it was just on the CB band instead of AM or FM.

We had a lot of fun with KPSR. At one point, Kenny and I got into some rivalry about whether KPSR was the station at my house or the station at his house. Then we decided to settle the dispute by making up a schedule where each of us put KPSR on the air at different times.

My parents were kind of neutral about the whole thing. They thought it was great that we were playing with radios and CBs and Lafayette Broadcasters, and not out in the street getting into trouble. But I'll tell you this — I wanted MORE POWER! I realized that if there was going to be any serious broadcasting, there would have to be a LOT more power than a Lafayette Broadcaster or a walkie-talkie could ever provide. And I was out to get it.

Chapter Two
I Want a Radio Station!

In the fall of 1965, I began my first year of high school at Lincoln High in Yonkers, New York. Lincoln is a predominantly white, Italian-Jewish high school that was built in the late 1950s as a model school for students in the 7th through 12th grades.

High school was a big change for me. It was a real dictatorship. You had no rights and you were always being told what to do. Because of that, I had a very healthy distaste for the whole high school system. I know there are a lot of people who enjoy high school, and some of the stuff we learned I found fascinating — especially science, math, social studies, and world history. But the attitudes of most of the teachers and the way they treated young people was just awful. And if there's anything that I don't like in my life, it's being told what to do. I guess it's pretty obvious in everything that's happened to me. But I just don't like it.

On the first day of school in social studies class, the teacher stood up and said, "I want to get one thing clear. We're going to be talking about world history and government and things like that — but this is NOT a democracy. This class is a dictatorship and I am the dictator!" I knew that I had to find a way to get transferred out of that class in a hurry.

Some of my classes in high school had wonderful instructors that made learning interesting, vibrant, and fun. And I always looked forward to attending them. But those kind of teachers were in the absolute minority. Most of the teachers were crazy, bitter old farts who made sitting in those hard seats all day long unbearable.

But it was those endless hours of sitting in hard seats in high school that made radio broadcasting and using radio as a method to touch people's lives and solve some of the world's problems even more attractive. I was always one that believed in Marshall McLuhan's idea of a global village. In other words, radio and other forms of mass communication can and should be used to connect us all together and make life a little bit more pleasant.

So as I sat there waiting for the final bell to ring and free us from school for the day, I often dreamed about being involved in the radio industry.

Of course, television is also an interesting medium. But technically, television is a lot more expensive to get into than radio. Back in the late 1960s, you just couldn't go out and buy video equipment the way you can now. At that time, a black-and-white television camera cost about a thousand dollars, and a color television camera cost $40,000-$50,000 — and that was in late 1960s money.

During my high school years, my friends and I were always involved in playing around with something electronic. There were three electronic whizzes at Lincoln High School — Albert Langella, Michael Kusiak, and myself. We were thought of as those weird kids who could fix anything, build anything, and do anything. Everybody looked at us kind of strangely, but at least they left us alone. Actually, a lot of the teachers and students thought I might blow up the school some day, so they were pretty nice to me.

Before the idea of radio broadcasting hit, my friends and I had a blast making home movies. We got together on the weekends to make films with an 8mm movie camera that I borrowed from my father. We made all kinds of movies — spy movies, science fiction movies, blowing-up-the-world movies. I even made a movie about a radio station first going on the air. Some of them turned into real productions.

Friday at school, we all got together to plan what we were going to do that weekend. Eventually, it got to the point that we'd say, "Well, what kind of a movie are we going to make?" My friends

Vincent, Jeffrey, and I would come up with an idea, put our pennies together, and spend $2.50 for a 4-minute roll of 8mm movie film. Then we'd plan everything, stage it, rehearse it, take a shoot, and send the film out to be developed for $1.50.

When the film came back, we all sat down to figure out which segments we wanted to keep and what we wanted to cut out. I had this really cheap film editor I used to splice it with. When the film was ready, we'd invite all our friends in the neighborhood down to my basement and have a viewing. Of course, all our movies were silent films. There were a few 8mm sound movie cameras on the market at the time, but they were way too expensive for our budget — and developing sound film was also very expensive. Every once in a while, we used a tape recorder to try and synchronize the sound, but it didn't work out all that well. So we just had to deal with making movies in the pre-talkie era. But still, it was a lot of fun.

I was always the kid who got to run the movie projector in high school. It all started when I was in 7th grade. Since I was interested in anything electronic, I sat close to the projector so I could try and figure out how it worked. Then one day the sound died, and the teacher didn't have a clue as to what to do. I noticed that the sound bulb had burnt out, so I walked over and said, "You'll have to remove the cover and replace the light bulb." She took the cover off and sure enough, the sound bulb had died. She found a replacement in the parts box, popped it in, and it worked. From that point on, I was put in charge of running the projector.

After that, I was quickly enlisted in the high school's audio-visual squad. Joining the AV squad was one of the best things I ever did in high school. We were excused from classes a lot. And it got me into movies, concerts, and all kinds of great performances. Sometimes I was pulled out of classes for the entire morning or afternoon to work at the rehearsals.

The Lincoln High auditorium was a spectacular place to work in. It had what is known as a light gallery, which is a huge cage suspended from the ceiling. They had spotlights up there as well as an audio mixing console and high-powered amplifiers. It was a lot of fun to operate.

On Friday mornings, the whole school was forced to go to assembly in the auditorium. All the boys had to wear jackets and ties — except of course, the guys on the audio-visual squad. The view up in the light gallery was great. And every now and then when the mood struck us, we could throw things down at people in the audience.

We also had a wonderful thing known as audio-visual identification cards, or AV cards. In high school, you could get in a lot of trouble for walking in the hallways during classes without permission if the Gestapo-style regime caught you. But if you were on the audio-visual squad, you just took out the AV card and said, "Look, I'm on audio-visual duty," and that took care of everything.

With high school being the dictatorship that it was, I naturally looked forward to summer vacation. The week or two before summer break was almost euphoric for me. The temperatures were going up, the sun was staying up longer, and the nights were heavy with that great smell of spring. I just counted the days until summer break.

Many lazy summer days were spent in the basement at 245 Kneeland Avenue tinkering with radios and listening to the various AM and shortwave stations that came in at night. Often, my vacation days consisted of getting up in the morning, going down into the basement, and playing around with my radio equipment. During the summer months, my goal was to try and get at least one radio to work per day. I also had a garden out in the back where I raised marigolds, tulips, and things like that. Around noontime, I usually went outside to visit my friends.

Then about 3 PM, I went out on the screened-in back porch of our house and listened to the radio. I usually tuned in WINS on the great old Zenith console our housekeeper gave me to fix up. At the time, WINS had an all-news format. I liked to listen to the news stories and the information about what was going on in the world. At 12 or 13 years old, most of it didn't really sink in. But still, I found it fascinating. This is the wonder of radio. Unlike any other medium, radio paints a picture in the listener's mind. And that is what I've always found to be absolutely fascinating about it.

A very pivotal point in my life took place when I was about 14 years old. One evening after school, my parents told me that there was going to be a documentary about the history of the broadcast industry on our local NBC television station.

14 years old at camp.

The program was great! It showed how influential radio was in the years before television was invented. People came home from their job at the office or factory, or came in from the fields after a day of work on the farm, and the whole family huddled around the radio set to listen to the news. Then during the evening, all the comedy shows, mysteries, dramas and kids' programs — *Milton*

Berle, Red Skelton, CBS Mystery Theater, The Shadow, and *The Lone Ranger* — came across the airwaves.

As I watched, I thought, "This is amazing! People would actually sit in their living room around their old Atwater Kent, Stromberg-Carlson, or Philco radio in the dial glow from the No. 47 pilot light and listen to stories that painted pictures in their minds." I found the whole idea to be very captivating.

The documentary also had some film footage of the early transmitting and studio equipment. At that point, I was really into anything that had to do with radio and electronics, so seeing all these great old vacuum tube transmitters, carbon microphones, and early wind-up turntables was very exciting.

That special on television was the turning point for me. When I saw how people were really captivated by the radio in the '20s, '30s, and '40s — not to mention what radio stations looked like technically — that was it. I was hooked. I knew for sure that I wanted to be a radio broadcaster from that point forward.

I said to myself, "I want to build one of these radio stations that people will crowd around at night and listen to for news, information, and stories. I want to DO this!"

That's what really started it. It was fun to play with the Lafayette Broadcasters, but now I wanted the big time. I wanted a REAL radio station. So I read and I studied and I studied and I read.

I went to the library and I started taking out every single book that I could find on radio broadcasting, equipment, transmitters, and whatever. I took out *A History of Broadcasting in the United States* by Erik Barnouw. It came in three volumes. The first volume was *A Tower of Babble*. It told of the first experiments of Oliver Lodge and Marconi in the late 1800s and the early years of radio broadcasting. It went up to the early '30s — right before the great radio networks came into being.

The second volume, *The Golden Web*, told about radio in the mid 30s and beyond. The third volume, *The Image Empire*, told the story of television's development. (Later on in my life, I taught a college course on the history of broadcasting in the United States, and used those three volumes as the textbooks.)

After I checked those books out of the library, I couldn't wait to get home from school and finish my homework so I could crack them open and soak up every word. And as I read them, a picture of radio broadcasting was painted in my mind.

I found out that radio was spawned by all kinds of people — inventors (some of them crackpots), a lot of make-a-buck Charleys, entrepreneurs, the whole gambit. And it was absolutely riveting. There were stories of the very first radio stations that were run by radio amateurs who were experimenting with voice modulation equipment. They put a carbon microphone in the antenna circuit to modulate the RF (radio frequency) and played phonograph records over the air at night — and people LOVED it!

Of course, KDKA in Pittsburgh, Pennsylvania, is the first station that was officially licensed by the U.S. government for broadcasting. But people had been talking and playing music over the ether since Reginald Fessedend's experimental Brant Rock, Massachusetts, station on Christmas Eve, 1906. Wireless operators on ships were astounded to hear music instead of the usual Morse code and static crackle coming over their headsets. This was the beginning of broadcasting — in other words, sending out radio signals to a broad area of people instead of just point-to-point.

So at age 14, I realized that radio really was, and in many ways still is, the loudspeaker of society. Granted, by the mid-1960s, radio had gone mostly to the disk jockey format and television was the dominant medium most people tuned in at night. But still, radio is what really touches a person, because it paints pictures in your mind and it stimulates your brain. You're not just sitting there watching images as you do with television. You're thinking. You're hearing a story and you're imagining: what does it look like? What are the faces of the people behind the story? That happens with radio whether it's a news story, a comedy, or a drama.

As I read about this wonderful way of communicating, I kept wondering, "How do I do it? How can I make it happen?"

When I asked my father, he said, "Building a radio station costs millions of dollars, son, so you might as well forget it. Just get yourself a CB set and talk on it."

When I talked to the kids at school, I got about the same response. They'd say things like, "Why don't you just become a ham operator?"

I said, "Nope, I want to talk to people. I don't want to just talk point-to-point. I want to talk to a broad audience."

That's what a 14-year-old kid was faced with. And I really could not comprehend why it was so difficult or expensive to put a radio station on the air.

I went to the library and read up on some of the Federal Communications Commission regulations. At one point, I contacted someone at the FCC field office in New York City and was told, "Well, if you want to do radio, you need about a million dollars." To me, that didn't seem right. Why should it cost a million dollars to do such a simple, basic thing?

So I kept thinking about it and toying around with the idea. Very early in my decision process, I came to the conclusion that the only way to get on the air was to build or buy a transmitter and go on the air myself — without the FCC's permission.

By that time, I had it in my head that broadcasting was just too good of a thing to need a license for. The whole concept of licensing didn't even faze me. Putting a radio station on the air was such a wonderful, benign thing to do that the idea of having to pay hundreds of thousands of dollars for equipment, buildings, property, engineering studies, and so on seemed like total nonsense.

I talked about the idea of setting up a radio station with my friends. Some of them were very supportive. But a lot of people thought I was nuts. They said, "Well, how are you going to build it... how are you going to do this and that?" And of course, this was still during the Lafayette Broadcaster days. But Lafayette Broadcasters were just a toy — the signal only went a few houses down the block. I wanted to build a real radio station that could cover the city of Yonkers — which had no local radio station of its own.[1]

One of the main problems I had to deal with in my early radio career was the lack of equipment. I had all these great ideas about putting up transmitters and antennas, but no equipment to make it happen. I started pricing some of the stuff you need to build a radio station, and it was outrageously expensive. I knew that I couldn't just go out on the open market and buy what I needed new, because it cost thousands and thousands of dollars. And I had no idea of where to buy it used or at a discount.

A great breakthrough in the process of electronic acquisition took place when I discovered the wonderful world of government surplus. One day, Michael Kusiak said, "If you want to buy parts and transmitters cheap, you should go down to Canal Street in New York City."

I asked, "What's that?"

"It's a place that has lots of stores that sell government surplus electronics," Michael told me.

So when my father came home that evening, I asked if he would drive me down there. Being a wonderful man who was always supportive of his children, he said, "Sure — when do you want to go?"

We decided to go the next day after school. And when we got there, I was like the proverbial kid in the candy store. I jumped out of the car and could not believe what I saw. Words could not describe the absolute awe and amazement of stepping into those stores and seeing floor-to-ceiling shelves and racks filled with the most wondrous electronic equipment of the time.

The old military surplus World War II stuff they had was big, black, and loaded with tubes, dials, and switches — just the kind of

---

[1] Yonkers, New York, a city of 200,000 souls, to this day does not have its own local radio station. In the early days of radio there were licenses issued in Yonkers on both the AM and FM bands, but they were later moved or canceled. During our pirate days in the late '60s and early '70s, one of our main reasons for broadcasting was that Yonkers had no local radio station of its own.

stuff you'd see in a Frankenstein movie. There were tons of it. And best of all, it was cheap.

I went looking through the stores, and it was just incredible — transmitters, receivers, test sets, radar equipment, antennas, all kinds of vacuum tubes — anything electronic that you could possibly want.

At this time I had about a dollar a week allowance, but my father was always very generous with me, especially if he saw that something was titillating my fancy. So he said, "Allright, you can buy something, but it can only cost about $10 or so."

To me, that was a fortune.

I decided to buy a Model TBS Field Backpack Set transmitter, built in 1935, which was priced at $9.95. If I had known how to dicker down the price, I probably could have gotten it for $5.95. But that's when the flood gates opened and I realized that getting parts and equipment to build my dream radio station was a possibility. If you could buy a 3-watt transmitter for $9.95, a 100-watt trans-mitter couldn't be that much more!

Once I discovered Canal Street, I kept asking my father to take me down to buy parts, equipment, radar gear, and other stuff to tinker with. But eventually, he got tired of having to drive all the way down there and wait for hours while I went through all the stuff at the surplus stores. That's when he decided that, at the age of 14, I was old enough to go down to lower Manhattan with my friends on the subway. This was back in the late 1960s when the subway system was safe. My father had no problem with letting me go there, but my mother, who was super-protective, was very concerned about it. But the decision had been made, and I was allowed to go.

My first impression of the subway was total awe. The immense underground system of tunnels and switches and lights was tremendously fascinating. We had no trouble getting down to Canal Street, and had absolutely no problems with anyone.

My friends and I went down to the surplus stores every other weekend or so. I found a lot of parts and transmitting gear, but nothing that was suitable for broadcasting on the upper end of the AM broadcast band. I wanted a station so badly that I tried to piece it together part by part, and it was still adding up to a pretty expensive outlay.

Then one Saturday when my father had time to drive Jeffrey and me down Canal Street, I found a piece of equipment that brought my dreams of broadcasting to a wide audience of people a lot closer to reality — a 100-watt marine radio-telephone transmitter. Now the nice thing about this transmitter was that it was self-contained and worked on standard household current, so you could just plug it into the wall outlet. A lot of military surplus equipment uses weird voltages — you have to either build big power supplies for them or run them off 400-cycle AC or some other strange configuration. But this particular marine set worked on 115 volts and transmitted on the shortwave band. It was tunable from 2 to 6 MHz. And it looked like something I could use to put a real radio station on the air.

It cost only $20, and we took it. It was heavy too — it weighed about 100 pounds! My father and I grappled with the transmitter, threw it in the trunk of his Cadillac, and brought it back to our house in Yonkers.

I immediately went to work on it and opened it up. It was a crystal-controlled rig. I was able to find some crystals that operated on the 80-meter shortwave band. I stuck them in there, screwed around with it, and to my amazement, I actually got the thing to go on the air. I think it put out a signal on 2.5 MHz or thereabouts.

I called Jeffrey and Vincent and said, "I have a transmitter that actually works! We need to build an antenna system, and then we can go on the air."

I was 14 or 15 years old at the time, and I really wasn't too aware of what I was doing from a technical standpoint. I'd read a lot about how radio transmitters operate, so I knew I had a lot of work to do. I had to put up an antenna tower, I had to tune it, and I had to load it. But unfortunately, I didn't have anybody to help me. I knew a couple of ham radio operators from school at the time, but they didn't want to get involved. When I told them I wanted to put a radio station on the air, they said, "Nope, we don't want anything to do with it!" So I got no help, no support, no nothing from them. All the ham radio operators I knew in high school were a bunch of snobs. That's probably what turned me off to the wonderful world of ham radio. Mind you, I think ham radio is great. And I have a lot of respect for ham radio operators who experiment with the medium and explore new and different techniques to further the communication arts. But the hams that I knew in high school were nasty, abusive, and very cliquish. In other words, if you didn't belong to their little group, to hell with you. So I was on my own in my efforts to figure out everything, read up on everything, and find a way to make it work.

So there I was with this shortwave marine radiotelephone transmitter. The transmitting equipment could put out 100 watts, so we figured it would go a long distance. And we had lots of shortwave radios to pick it up on. My father had a couple of short-wave receivers, and some of my friends had Lafayette Explorers or other inexpensive shortwave radios. Nearly all of my friends went through the stage of buying a shortwave radio kit or buying a ready-made shortwave radio receiver. We were all into that.

My friends and I spent a weekend or so playing with the transmitter, trying to get some power out of it. Jeffrey and I erected an antenna out of pieces of pipe, wire, and whatever stray pieces of metal I could find laying around the house. Then, much to my father's chagrin, we decided to strap the antenna to the banister of the upstairs porch. I ran a piece of wire down to the transmitter, hooked it up, turned it on, and fed it some audio. I hooked up a carbon microphone and switched everything on. And to my amazement, sound came through on my portable shortwave radio on 2.5 MHz. Everybody was delighted.

Now that we had the transmitter working, we finished assembling our station. For our studio, we had a phonograph and a tape recorder that I borrowed from my father. We had an old chrome carbon communications microphone to speak into and a carbon microphone set in front of the turntable's speaker so we could play records over the air.

When I was finished with the station layout, everything went through a switchbox I built inside of a wooden cigar box. You could either switch on the microphone that was in front of the phonograph and play recorded music, or you could switch over to the other microphone and speak over the air.

This stuff was real bare-bones equipment. The phonograph was something I found in the garbage. It was hooked up to this horrible one-tube amplifier that barely gave out a recognizable signal. The transmitter's audio quality was worse than a telephone line. But back in those days, we didn't even think about audio quality. We were happy just to be in the air.

Once the station was finished, Jeffrey, Vincent, and I started to discuss what we would do for our very first program, which took place in the summer of 1968. When the big day finally came, I switched on the transmitter and Jeffrey said, "This is WRAD Radio, and now we're going to hear..." I don't remember what the first piece of music was, but it was something from my parents' collection of popular and classical records.

While Jeffrey was on the air doing his program, Vincent and I ran around the neighborhood with a portable radio doing field tests. We soon found out that the signal went all over the neighborhood. My friend, Kenny Sofer, picked it up across the street, and so did Larry Rand, who lived a few blocks away.

Considering that we weren't using anything for a transmission line, the grounding system wasn't all that great, and the transmitter probably wasn't tuned correctly, I would say it did pretty well. On 2.5 MHz or thereabouts, WRAD was putting a signal out that could be heard for about a mile or two. And that to us was very, very exciting.

For most of that summer, Jeffrey came down to the basement, sat in front of that microphone, and played 45 and 33 RPM records for a couple of hours every day. And he loved it. Vincent and I walked around the neighborhood to see just how far the signal went. Our friends called us up on the phone and asked us to play musical requests. All of our friends who had shortwave radios tuned us in. To my knowledge, I don't think we ever had any outside listeners. We only operated our station during the day. I never thought of operating at night — but if we had, we probably would have gotten some skip coverage into other areas.

WRAD was our first broadcast station above the Lafayette Broadcaster stage. We even used to keep a log — it's probably buried somewhere up at my farm in Monticello, Maine, where I keep my archives. Jeffrey did most of the programming, which was playing 45s and 33s, talking into the microphone, and reading news. It ran like that for the summer, and it was fun. But after a while, I started trying to figure out a way to have a station on the broadcast-band frequency where more people would hear us.

In the winter of 1968/1969, Michael Kusiak gave me a catalog from Fair Radio Sales Company. They are a big electronic surplus house, catering to amateur radio operators and electronics buffs. When I opened up the catalog, I just went crazy. They had all kinds of great stuff — radar equipment, transmitters, receivers, everything! And in the back pages of the catalog was an advertisement for a brand-spankin'-new World War II BC1100 military transmitter.

A transmitter like that would be just perfect for our station. It worked on 115 volts AC, so there weren't any crazy, expensive power supplies to make or buy. Secondly, it covered from 1.5 to about 8 MHz, which meant that it covered the upper end of the AM broadcast band. Thirdly, it ran about 50 to 100 watts, so it would put out a signal you could hear all over Yonkers. The price was $125. But it might as well have been a million dollars, because for someone who was on a relatively fixed allowance, that was an unattainable amount of money to raise.

Still, I wanted it. I showed the catalog to my friends and told them this was the transmitter we needed — because to build something of equal power and modulation would cost more in parts alone than the complete transmitter. One day after school I was reading about my dream transmitter in the catalog when Mom came into the room. I showed her the picture and said, "Look — I want this transmitter."

A transmitter might as well have been a toaster to her, but she looked at the picture and said, "Okay, I'll talk to your father." It was always like that with Mom. Whenever I wanted something or wanted to do something, it was, "Well, I'll talk to your father about it."

Then all of a sudden, she came to me and said, "If you want to order it, go right ahead." I was flabbergasted. I couldn't believe it! At 15 years old, $125 was a lot of money.

I took out the order form and wrote down, "One BC1100 transmitter." My mother wrote me a check for $125. I dutifully put the order in the mail and off it went. Even before the mailman picked it up, I was counting the seconds to the time when the first REAL transmitter would arrive at 245 Kneeland Avenue.

The picture in the Fair Radio Sales catalog showed what looked like a table-top piece of equipment — in other words, something you could put on top of a table. Well, about two weeks later as I was arriving home from school, my mother walked out of the house and said, "It came." I was all excited. Then she said, "It's in the back yard."

I said to myself, "In the back yard...WHY?" I walked down the driveway alongside of the house and here was this gigantic crate sitting in the back yard. Now mind you, as a 15-year-old, probably the biggest thing you normally move around is school books. So being confronted with this huge crate sitting in the middle of the back yard was something. I walked up to it and gave it a little shove but it didn't move at all.

My mother, being the good sport that she was, said, "Let me get on the phone and get your cousins and your friends and everybody over here." The crate was very well constructed — it was screwed together and not nailed. So while my mother was on the phone trying to enlist an army of relatives and friends, I found a screwdriver and started working on it. There must have been a hundred screws on the bloody thing. It took forever for me to get the top off. And then, there it was — all 400 pounds of it — this gigantic, gray-painted, brand-new military surplus AM transmitter. My cousins, Jacky and Edward, came over, as well as Kenny Sofer and Larry Rand.

Believe it or not, the thing had handles on it. That's what I like about military surplus equipment. If something weighs hundreds of pounds, they have enough sense to put handles on it. So everybody grabbed a handle — and with the help of about six people, we lifted it out of the crate and grappled with it down the basement stairs. Getting it down the basement stairs was a real challenge, because not everybody could fit on the stairway at one time. At one point the transmitter was kind of teetering on the edge, but we finally got it down to the basement. We had a dolly, so we set the transmitter on it and rolled it into the main part of the room. The manual said that the crate it came in could be used as a stand, so we put the crate in a corner of the basement. Then all six of us grasped the transmitter and hoisted it up on the crate. My mother gave everybody lemonade, and I thanked them for coming over to help.

The BC1100 had that intoxicating odor of tropicalization — this lacquer-type stuff they put on military equipment so it would resist fungus and corrosion. It was brand-new — it had never been used. The tubes were packed separately. I took the manual, skimmed through it, and put all the tubes into their sockets. That evening, I realized that I had a lot of work to do before we could go on the air.

First, I had to run power lines to it and figure out what to do for an antenna system. So for the next two or three days, I was busy hooking it all up. I built the antenna system, which was a 20-foot whip, and lashed it to the sun porch of our house using rubber tubing for insulation. I didn't have enough transmission line, so I had to splice together about five pieces of whatever I could find. I didn't even care what kind of cable it was — RG 8, RG 11, RG 59. I just spliced it all together. And I didn't have any electrical tape either. The only thing I had available was masking tape, so that's what I used.

Next, I scrounged up some power cable. I found bits and pieces of it and spliced them together. I ran a separate power cable from our breaker box to the transmitter, and I used a big knife switch as the main disconnect. But I still had one big problem. I didn't have enough power cable to run from the knife switch over to the transmitter.

I had most of the wiring hooked up and I was just about to the point where I could turn on the transmitter and test it, but I had to go and work on the audio-visual squad at the high school. When I was up in the light gallery, I happened to notice a coil of power cable laying over in the corner.

I asked the teacher, "Can I have this? I need it to hook up the transmitter."

And he said, "I'll tell you what. I'll make you a deal. You stay here until 10 PM tonight and help me run these spotlights, and I'll give you the cable." Originally, I was supposed to stay for only an hour. I really wanted to get back and work on the transmitter. But I did need that power cable.

So I said, "All right," and stayed there until 10 PM, counting the minutes and seconds until I was free to go back and work on my transmitter. When 10 o'clock finally came and the performance was over, I grabbed the cable and ran home.

By then, it was already late and my parents insisted that I go to bed because the next day was another school day. But I really wanted to get the power cable hooked up. Before I went to sleep that night, I wanted to throw a switch and see if the transmitter would come on.

With military electronics, even though a piece of equipment is new it doesn't necessarily mean that it will work. The transmitter had been in storage since 1945. And after all that time, who knew if it would work? So I hurriedly hooked up the power cable.

But by then it was past 11:00 PM, and my mother was just livid about it. She said, "That's it... you've got to get to sleep." So I connected the power cable, got voltage into the transmitter, and very grudgingly went upstairs to go to bed.

The next day, school was excruciatingly long. When the last bell finally rang, I ran home and threw on the power switches. Then I got out the maintenance manual that came with the transmitter and carefully followed the directions. It said to flip this switch. I did, and a light came on. So at least that part was working.

The BC1100 was an interesting transmitter. It was designed to be operated by remote control over a telephone line, and there was an actual telephone dial on the front of it. To operate the thing, you first dialed the filaments up. Then you dialed the preset channel you wanted to use — A, B, C, or D. After that, you dialed whether you wanted to transmit with phone (voice) or Morse Code. That's how it worked — it was really neat!

Using a 100-watt light bulb as a dummy load, I dialed the filament on and the whole thing came to life. All the pilot lights went on, the filament voltage meter went on, the blowers went on, and the tubes lit up. I was ecstatic. I tuned to channel A, put it on phone, and turned the plate current on. After about five minutes of tuning, the dummy load light bulb lit. By then, I was just going wild. This was such a joy. This meant REAL radio!

The next thing to do was to tune it up to the antenna. We had an antenna current meter on the transmitter. The biggest problem was finding a frequency to tune it up on. Since I didn't have a frequency counter, I had to find a blank spot on my father's Tarryton transistor radio. In New York City, we had an AM station — WWRL on 1600. So I tuned the radio to WWRL and pushed it up past them a crunch. Then I tuned the transmitter's oscillator until I could hear it on that spot. That's how I set the frequency.

Albert Langella is the one who suggested we go a little bit above the standard AM band. (We thought of using the "expanded AM band" before the FCC did!) I had always been concerned about causing interference if we went on the AM band. But if we were above it, our signal wouldn't bother anybody.

So I called up Albert one afternoon and said, "Look, I think I got the transmitter more or less on the air. Try to tune it in. I think I'm putting out a signal a little bit above 1600."

He put the phone down and went to check his radio. Then he came running back yelling, "You're there! You're there! I got ya! I got ya!"

Then Albert said, "But you're a little too low. Raise it up a crunch." So I went over to the local oscillator dial and just tapped it. I could hear him yelling from the other room, "That's okay — right there." He came back to the phone and said, "Well, you did it — you're clear! You're all right where you are. You're a little bit above 1600 and you're not interfering with anyone."

I was in ecstasy. Albert lived four or five miles away, and he was hearing it great. I said, "Well, if you can pick it up, that means everybody in Yonkers can pick it up!"

Following that were days of walking around with transistor radios and listening to tones. It's got to be the right of any pirate radio operator who puts a transmitter on the air to walk around or bicycle around the neighborhood and listen to their tones. And that's what we all did. I estimated that we had about a 10-mile range. It wasn't a lambasting signal — that's for sure. But it put a good 50 watts into that weird antenna system I had strapped around the porch.

We had a real low-budget operation there. But despite its craziness, it worked! I grounded the transmitter to the cold water pipe in the basement. I ran a special line for that because I knew that grounding was important at medium-wave frequencies.

I realized that I had to build some kind of a studio for our station. For a kid my age, finding audio equipment meant scraping up whatever I could get from my friends and pilfering whatever I could steal out of my father's stereo system. And that's just what I did. I took the tape recorder from the stereo and borrowed one from a friend. I borrowed the old speakers that my father had used on the hi-fi system upstairs in the TV room. My father had a couple of microphones that came with his tape recorder, and I used them as the main microphones.

For the audio console, I took a cigar box and put a whole bunch of toggle switches on it. It was that bad — no pots, no attenuators, no correct matching — just all hard switches. And did I have a VU (volume units) meter? Forget that! All I had were these junky meters that I'd salvaged from surplus electronics equipment. Whatever meters I had I could use a diode to rectify could indicate audio. The transmitter came with a remote control unit that had an amplifier in it, so I used that as our line amplifier.

I must have spent the whole weekend hooking everything up. One of the biggest problems I had was figuring out how to match our audio equipment to the 600 ohm lines, which is what the transmitter worked on. Once I was finished, I was able to pipe audio through this transmitter directly. In other words, I didn't have to use the microphone-in-front-of-the-speaker trick. The audio quality was listenable, but it was poor. It was the kind of audio quality that could give AM radio a bad name. But to us, it was wonderful to be able to put any kind of radio transmission on the air.

We set up the audio console switchbox on a table in the basement. We had a turntable on one side and a tape recorder on the other. Our record collection was what I could get out of my parents' collection, which was classical, beautiful music, and a lot of Herb Alpert and the Tijuana Brass. Pretty pitiful, but that was it.

My friends — Jeffrey, Vincent, and Albert — started planning when we were going to go on the air with the first big broadcast. As teenagers, we really put a lot of thought into this, because we figured that if we went on the air, the second we turned on the transmitter, the whole world would be listening. Actually, I was more realistic than that, but I was hoping we'd have plenty of listeners for our first show. But just to make sure nobody missed it, we figured that we should do some advertising.

Like nearly all teenagers, our world revolved around high school. Our world was getting up in the morning, getting dressed, going to school, and counting the seconds until the last bell rang and we were free to go home or have fun with our friends for the rest of the day.

So the high school seemed to be the natural place to advertise our station. We decided to type up some leaflets and hang them up all over the school. The leaflets said:

STARTING AT 4 PM ON (some date in October)
A NEW AND DIFFERENT KIND OF
RADIO BROADCASTING WILL BE PRESENTED.

I typed the leaflets up on my Underwood No. 6 manual typewriter using carbon paper. Back then, copying machines were difficult to get hold of. There was a copying machine at my father's law office, but we thought that to use it would be a security breach. My parents knew that we were doing something, but they didn't know all the details. I was afraid that if my father knew exactly what we were doing, he might not agree with it.

Actually, my parents turned out to be pretty neutral about the whole thing. They always were neutral. My mother knew that we were doing something, but she thought I was better off in the basement playing with radio equipment than out on the street getting in trouble, fighting, getting drunk, or mixed up with drugs like a lot of kids at the time were doing.

You know how parents are — they know you are doing something but they just don't want to acknowledge it. They just change their reality and say, "Well Allan is in the basement, so at least he's home." After all, what possible trouble could a kid get into when he is down in the basement playing around with all that crazy radio equipment?

At the time, we didn't worry too much about the legal implications of our station. The idea of being "pirates" hadn't even entered our minds. We weren't interfering with anyone, so we figured that what we were doing was only a few steps above the legality of CB radio. But still, we decided to give ourselves code names. I called myself "The General," Jeffrey called himself "The Chief," and Vincent called himself "The Professor."

We put the leaflets up all over the school and started to plan our first big broadcast. We talked about it every chance we had.

Vincent, Jeffrey and I decided to go on for an hour — from 4 to 5 PM. We set our starting time at 4 PM because we needed time to get home from school, tune up the equipment and get ready. We decided that for our first broadcast, we would play a little bit of music and talk to anyone who was listening out in the community.

People were very curious about the leaflets we'd put up. Everywhere we went at school, people were talking about our station. Even the teachers were wondering what it was all about — so we knew we would have people listening when we signed on the air.

I went home that Monday afternoon, immediately warmed up the transmitter, got the plate on, and put the carrier on the air. When 4 o'clock came, we hit the plate and away we went. We played a few selections from Herb Alpert and the Tijuana Brass, and then I got on the microphone and said, "This is WRAD, just a little bit above 1600 on the AM dial." (We had no idea whatsoever what frequency we were on. We could have been on 1620, we could have been on 1623, or even 1630.) I introduced myself — Jeffrey and Vincent were there and we all shared one carbon microphone. We played a little music and then I turned it over to Jeffrey, who from the earlier shortwave days of WRAD was very good at playing records and announcing.

I took the portable radio and went walking all over the place. And sure enough, all my friends came running outside yelling, "Hey, Allan, I picked you up! I got you on every AM radio in the house!" Everybody was picking us up! And the next day in school, people were coming up to us and saying, "Hey — we heard you!" and, "It came in loud and clear!"

But then I starting hearing rumors that the principal was listening and he was annoyed about it. Some of the teachers had also heard it and were wondering about what we were doing, and if it was legal. I started to get a little bit concerned, but then I thought, "We're only on for an hour a day, so what could possibly go wrong?"

By the third day, the paranoia was steadily building. We heard that the ham radio operators at school did not like what we were doing. And there was really no reason for anyone to complain. The programming at that time was very mild — we aired light music and weather, talked about things that were going on at the high school, and so on. There was no cursing. We were just getting our feet wet. But to be on the safe side, I told Jeffrey and Vincent that we should establish lookouts at least once every 15 minutes and have someone look around the house, just in case any of the hams from school came over to cause trouble.

Jeffrey was on the air and Vincent and I were hanging around outside. As we had decided, we walked around the house about every 15 minutes to check things out. About 30 minutes into the program, I started to get a strong feeling that something was wrong. It was almost like a sixth sense. I walked around front, and sure enough, just as I turned the corner, I saw two full-grown men standing in the doorway. They both were wearing black trench coats, just like agents in the movies. They were in the process of ringing the door bell so they didn't see me. I said to myself, "Oh my God, the FCC!"

At top speed, I ran down the driveway and into the basement. The only words that came out of my mouth were, "Get the hell out of here, it's the FCC!" Vincent ran out so fast that he almost knocked me over. Jeffrey, who was slightly disabled, jumped out of the chair and started running full speed at the same second. I had no idea he could go that fast! The whole scene was so unreal that it looked almost like something out of a cartoon.

And seeing what they had done, I did the same thing. We all scattered in separate directions. I don't know where Jeffrey went. I think he ran all the way home, which was a good two miles. And I think Vincent did the same thing. I went to the reservoir, where I always liked to hang out as a teenager and stayed there for two hours or so. But finally, I decided to go back home and see what was happening.

I found out that my mother had answered the door, invited them in, and served them coffee and cake. They informed her that they believed her son was operating an illegal radio station. She took them down to the basement to look at our equipment, and they proceeded to remove all of the transmitting tubes and cut some wires. The agents then went on to inform my mother that our station was in violation of Section 301, and that my father would receive a letter regarding the matter in a few days. When my father came home and found out what happened, he said, "Well, son, you can't do that anymore." And he was pretty firm about it.

That was my first bust. I knew that some of the kids at school didn't like what we were doing, but I was shocked that the government came. Our signal was only going out five or ten miles, and we weren't doing any broadcasting at night. So I figured that one of the hams at school must had tipped them off.

The next day, I ran into my prime suspect. I went right up to him in the school lunch room, looked him straight in the eye and asked, "Did you call the FCC about my station?" and he said, "Yes, I did!" I was furious. We got into an altercation, and needless to say, it took a few people to pull us apart.

Let me put it to you this way. This guy was slime. He gave ham radio a bad name. And the only thing I didn't understand is why he did it. And all he would say is, "Well, you're violating the law and you can't do that"... and blah, blah, blah... I told him that I had just as much right to be on the air as he did, but he didn't buy it. For some time after that, the word "ham" meant THE ENEMY. I don't feel that way now. In fact, I have great respect for ham radio operators and what they have contributed to technology. But this particular guy, who shall remain nameless, was such a slimeball. He was nasty, evil, and was purposely doing this to create harm. Besides, no one likes a rat.

The interesting thing about that guy is, for a while, we were friends. We actually got along together and shared some electronic interests. That's one of the things that bothered me the most. I think he ratted on me out of jealousy, which is a pretty sad and sick reason for anyone to do something.

The next weekend, I promptly went down to the city and bought the exact tubes that the FCC took. I got some wire and my soldering iron and pieced the transmitter back together. In about 30 minutes after getting the tubes plugged into their socket, I had the transmitter back up and ready to go on the air. But I didn't do any actual broadcasting until later.

Chapter Three
WKOV Brings Free Radio
to Yonkers, New York

It was late 1969 in Lincoln Park with its tree-lined streets, English Tudor and colonial homes — a real suburban neighborhood. Well, deep in the bowels of the basement at 245 Kneeland Avenue was a tattered and knocked-off-the-air-after-three-broadcasts radio station, WRAD, broadcasting a little bit above 1600 on the AM radio dial.

The government had come, done their thing, and left. After the bust, I was naturally a little nervous about putting the station back on the air again. But I firmly believed that what we were doing with WRAD was right. And even after all that happened, I felt that there must be some way to get on the air and stay on the air.

After the bust, we changed our call letters to WWJ. We just broadcast for an hour here and an hour there, hoping not to attract the FCC's attention. But even with our limited schedule, my father was worried that there would be more visits from the government. There were also threats that school officials were going to turn us in to the FCC or have the police shut the station down if we didn't stop broadcasting. So from time to time, we went off the air for a while and just planned things.

In the closing months of 1969, WWJ didn't broadcast very much. But in the time it was on the air, the station incited quite a few people — including the principal of Lincoln High School. He didn't like the station at all, especially the nasty things we said about him.

1969 was a very turbulent time in America. There was quite a bit of unrest in the country, and people all across the nation were holding protests against the Vietnam War. Young, middle-aged, and old people marched in the streets demanding answers from their government. The Vietnam War was a direct influence on the times because it was such an insanity — a stupid, worthless waste of lives, money, and resources.

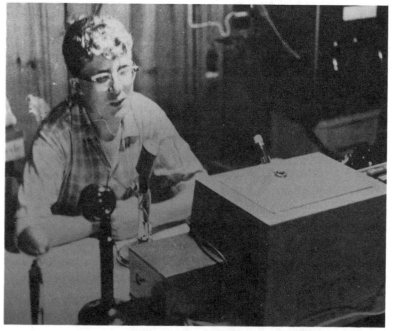

Jeffrey Slauenwhite at the controls of WWJ, 1969.

In the predominately white, Italian-Jewish neighborhood I was living in — an area called Lincoln Park, which bordered with places such as the Yonkers Raceway and the Bronx, it was relatively quiet. But even here, there were protests. Some of the seniors at Lincoln High School held demonstrations from time to time. They paraded around the main courtyard of the school protesting the war in Vietnam, the "pigs," the military-industrial complex, and the government in general. They had signs about flower power, love, peace, and all those wonderful platitudes that were being bandied about back in the late '60s.

While I was not directly involved with the protests, I was always curious about them. Up until that time, I was a relatively conservative person. I wasn't conservative to the point that I believed in everything the government authorities said. It was just that my main interests in life were electronics, science, and technology, and I wasn't all that concerned about the government and politics. In other words, I hadn't really "become politically aware" yet.

At that time, I was still in limbo about what I wanted to do with the station. I knew that I wanted to build a radio station, play music and say things on it, but I wasn't very clear as to exactly what I wanted to say. In some ways, I was hung up on technology. In other words, the nuts and bolts of building the station and the excitement of putting it on the air and hearing a signal come out of a transmitter of my own design or modification were my primary concerns. But now the political atmosphere had become so highly charged that it was definitely starting to shape the way I felt about things.

Aside from radio, 1969 was a tough year for me. My mom was in and out of the hospital. She had ovarian cancer, and apparently my father was the only one who knew. For some reason, he didn't want my mother to know exactly what was wrong with her, which I think was a mistake. And he kept her cancer a pretty good secret from me and my sister, Barbara, as well.

My mother had clinical depression for much of her life. It drove my father crazy and made life in the house hell. She would have terrible fits of depression and take it out on my dad and my sister. She was always good to me, though. I have no complaints. Ultimately, I suspect that the clinical depression was responsible for her death. If anything can be said about mind over body, my mother's depressed state could not have done any good for her system. And I think her cancer was directly related to that. She was a very depressed woman — very, very sad, and not knowing where to turn. Nobody knew what to do to help her. That was a time when clinical depression was not really recognized as a disease. Doctors were just beginning to realize what it was and experimenting with ways to treat it.

In December of 1969, my mother was admitted to the hospital for the last time. She was extremely sick. The cancer had spread throughout her body, and was literally consuming her alive. My dad took me to the hospital on January 1, 1970, but by then there really wasn't much to see. My mom was so riddled with cancer that the doctors had to keep her heavily drugged with morphine. She probably had no awareness of who she was or where she was.

At about 10:30 AM that day, she died. Mom and I had always been close, so naturally I was upset. My sister was mildly hysterical, and my father was extremely depressed. My mom was only 47 years old when she died. She was an extraordinarily beautiful woman, right up to those last moments. That was my first look at death. I was only 16 years old at the time.

The following year was a big adjustment period for me. It was definitely a very rebellious time in my life. Like I said, I hated high school. I hated the regimentation and I totally disliked the way I was treated. And I rebelled — but I rebelled in different ways than a lot of kids do.

Being a member of the audio-visual squad, I was able to obtain a copy of the high school's master key, which you could use to open just about any door in the high school. I made a soap impression of it and ground myself a copy. Once I had my own master key, I made visits to my high school at night and stocked up on supplies for the radio station. I took a lot of stupid stuff like spark coils, vacuum tubes, and things like that — nothing of any serious value.

But again, I was ratted out. One day in the spring of 1970, all these policemen came to my door with an arrest warrant. Then they went down in the basement and found all the stuff I had been taking out of the high school at night.

I was arrested. It was the first time I was ever arrested, and it was NOT for pirate radio — but for breaking into Lincoln High School and stealing things like tubes, spark coils, and all kinds of weird stuff. The experience was not fun. It was the first and last time I got involved in that kind of shit. I was treated as a youthful offender and the judge let me off with a slap on the wrist. I was also suspended from high school for three or four months.

The experience taught me a good lesson. What I did was wrong. It was interesting, though. The incident gave me a great reputation in high school. People thought of me as this mad Frito-Bandito scientist who built a radio station that was also busted. So when I returned to school after my suspension was over, the assistant principal called me into his office and he asked me how I was adjusting. I told him that I was doing fine. Then he said that a lot of people in the high school were really scared of me.

I asked, "Why would anyone be scared of me?"

And he said, "Oh you know, you were arrested for stealing some stuff and having the radio station the government closed down."

But my problems with the school authorities weren't over yet. Another time when I was in high school, I got sick and asked to be excused to go home. My parents couldn't be contacted and the principal wouldn't excuse me. So I just walked out. I was home trying to recuperate and I got this hysterical call from my father. He said that the high school called him and said I had walked out of school without being excused.

I was really pissed off at the principal. He knew that I was sick and had a fever.

The rain was pouring down that day. I put on my trench coat, walked up to the high school, and I barged into the principal's office. When I got there, he was in the middle of a meeting. But I was mad. I looked at him and I said, "If you EVER bother my father again, there will NOT be a school here the next day!" I really got nasty with him. And then I went home.

A little later, my father called me up and said, "Allan, the police are on their way. They think you're going to blow the school up! What's going on?"

I said, "Dad, come on... I was just angry." So my father, who has a wonderful way of solving problems and smoothing things out, called up his friends at the board of education and calmed things down. He smoothed things over and shut up the principal.

I was suspended again for a week or so. But this time, I had to go to the Department of Education and see the school psychiatrist before they'd let me back in. After talking with me all afternoon, she decided that I was just a normal teenager and that it would be safe for the principal to let me come back to school again.

I never got suspended for fighting or any of the traditional things that people got suspended for in school. But I was always getting suspended for something.

The summer and fall of 1969 was the peak of the "free" movement and the year of the Woodstock bit. It was also the year that some of my friends — especially Kenny Sofer — were becoming "liberated." They were doing the same thing that a lot of people in the late 1960s were doing. They were beginning to question what was going on.

Jeffrey, Vincent, and I had been going on the air from time to time with WWJ. We played Herb Alpert and the Tijuana Brass and things like that. We weren't really saying much, and neither was the music. But all that changed when, thanks to Kenny's influence, I became aware.

It all started one day when I saw Kenny Sofer and said, "Kenny, I have this radio station and I don't know what to do with it. I want to put it back on the air. There's all this stuff going on, all these people are trying to say things about the establishment... what can I do with this station?"

He gave me a whole bunch of Jimi Hendrix, Bob Dylan, the newly released Woodstock album that had just come out and said, "Well first, listen to this music." I remember going home later that night and putting all this music on the turntable down in the basement. And that was the beginning of the enlightenment of Allan H. Weiner.

After that, there was no more Herb Alpert and the Tijuana Brass. No sir! I was tuning in and turning on — mind you, no drugs. A lot of my friends were sneaking into the basements and back rooms of everyone else's house and trying marijuana, but I had a theory about drugs at that time. I really felt (and still do to this day) that anything that seriously alters your natural state of being is wrong. And I did not believe in pumping chemicals into your body because it destroys your ability to be who you really are. So even though a lot of my friends were getting into the drugs, smoking marijuana, tripping on acid, and doing all that stuff, I refused. I wanted nothing to do with it.

But in the summer and fall of 1969 during my enlightenment, I finally realized what our radio station could do. Kenny and I and some of his friends got together and decided that we should go back on the air. We would have political discussions about what was going on and we would talk about the unrest. Our radio station would be the loudspeaker for our friends at school and in the neighborhood. We would give a voice to people's feelings over the public airwaves. It was the rebirth of free radio in Yonkers, New York.

We decided that we needed a new set of call letters. Kenny came up with WKOV. They didn't really mean anything — they just sounded good. And they looked good when you wrote them down.

Kenny borrowed a picture of "Mr. Natural" with a microphone in his hand from *Zap* comics and used the silk-screen press he had in his basement to make posters for our station under our new call letters of WKOV.

I redid the studio a little bit. I set up a few extra microphones because I figured that since we were going to have panel discussions, more people would be coming down to the studio. We also had a music director — a guy by the name of Larry Nager. Larry had an immense music library — at least, bigger than anything that Kenny had. We all went over to Larry's apartment to discuss the music situation and decide how to program WKOV.

Larry volunteered to provide us with all the music we needed. I went to New York City and bought several 10- or 12-inch reels of used Muzak tape for $10. Then I bought a bunch of plastic take-up reels, wound the tape on that, and gave it to Larry to record his music on.

Once everything was ready, we put signs and posters up all over the high school announcing that on September 23, 1970, Radio WKOV would go on the air.

"This is WKOV radio... your natural station! 1610 KHz amplitude modulation." That's how it started. The first show was an hour long. It was me, Larry Nager's music, a lot of the Beatles, Hendrix stuff, and Bob Dylan. I think I talked about the peace movement and what was going on with the kids getting out in the streets, the protests against the war, and so on.

The first broadcast went smoothly and without incident. It ran only an hour, mostly because I was still pretty paranoid about the government and the hams from school causing trouble. I said, "Well, we'll go on for an hour every afternoon or so from 4:00 to 5:00, and that's it." But a lot of people were listening to us due to all the advertising at the high school, so eventually the broadcasts got longer.

One day in late September, Kenny brought over a whole bunch of people from a group called "Students Against the Vietnam War" or something like that. One of the members was a gal by the name of

Cheryl Beaver. She became involved with WKOV for quite a while. Cheryl was also one of the first people on whom I developed an incredible crush. It never really went anywhere, even though I guess it could have gone a lot of places. But I really fell in love with that gal.

It was a big broadcast for me because it was the first serious political discussion we had on WKOV. The group came down to the basement, sat down with Kenny, and started talking about the war in Vietnam. Cheryl Beaver had a prepared speech. She was talking about what all the "pigs" were doing and the corrupt government officials. Then she started using one four-letter word after another.

I had on my headphones and was watching the VU meters. And I kept saying to myself, "Okay, Al, this is free radio." At that time, I didn't know what to do about bad language. But I was slowly becoming more and more liberal. So I said to myself, "Hey, this is the language that people use, so it has every right to go over the air in good context." It was a very heavy discussion.

The broadcast was heard by the principal of the high school and some of the teachers. People even told me that it was heard by the Yonkers police department. It was a real hard-hitting talk. That one afternoon catapulted WKOV to the forefront of free political radio. It was exciting, it was fascinating, and it led to many more interesting broadcasts. It was an important day in the life and times of WKOV radio. The station was becoming very controversial.

WKOV was turning into a REAL radio station. Larry provided the music and we had all kinds of political discussions going on. But our operation was not continuous. Sometimes my father got very nervous about it or we heard some threats at school, and we decided to keep it off the air for a week or so. Then when things calmed down, we'd go back on the air again.

In the later part of 1970, another good friend of mine came into my life — a fellow by the name of Michael Schaitman. Michael grew up in the Bronx and was a friend of Kenny Sofer's. He didn't want to live with his parents full-time, so he hung out with Kenny and the people at the station. We found lodging for him in basements and spare bedrooms so he'd have a place to "crash." Michael and I became very close friends. Eventually, he moved into the basement of my house and he helped me run WKOV.

Michael and I had a lot of fun together. One of the most memorable things we did was climb up to the top of the Armstrong Tower. I grew up seeing that tower from Lincoln High School, and I always wanted to check it out. It looks like a big power tower, only much larger. It's a beautiful tower — 400 feet high with three 150-foot crossarms. At that time, the tower wasn't being used for FM broadcasting anymore — it was just used for 2-way mobile radio communication. And the old Edwin Armstrong field laboratory was still there.

Michael Schaitman

Michael Schaitman and I drove out to Alpine, NJ, on his motorcycle to see it one afternoon. The gate was open and there wasn't a soul around. So we decided it would be okay to check the radio tower out. We walked in, went right up to the tower, and proceeded to climb it. It had a steep stairway that you had to use to get to the top of it. We spent an hour or so just climbing the thing.

We finally got all of the way to the top, and we could see 100 miles around. We could see down to New Jersey, up to Connecticut — you could even see Manhattan beautifully. It was a wonderful experience. I could just imagine what fun Major Armstrong had swinging around that tower adjusting his antenna systems. He loved heights. We had a great time.

Allan H. Weiner on top of the Edwin Armstrong radio tower,
Alpine, New Jersey, 1971.

But then we looked down and saw police cars parked all around the base of the tower. Michael had already started to climb down, but I was still all the way up at the top. So when I saw those police cars I said to myself, "Well, maybe I'll stay up here for a while."

Michael told me that by the time he was back on the ground the cops were saying, "How are we going to get that fellow all the way at the top down?" And they looked at each other and said, "Well, I ain't going to go up there and get him!" Then one guy said, "Well, maybe we'll shoot him down, ha ha ha." I sat up there for about half an hour, but the police still hadn't left. So I decided that I'd better climb down and see what they wanted.

We were detained for trespassing. Actually, the Alpine, New Jersey, police were pretty nice about it. One of the older guys at the police station remembered when the Major used to live there and what a nice guy he was. We sat around and talked about the Major for a while.

Allan H. Weiner by radar dish, Armstrong Laboratory, 1971.

I told the officers that I'd called up the owner of the facility and asked for permission to walk around the area. They tried to call the owner, a real estate concern in New York City, but they weren't able to contact anyone. The police finally decided that we were just a couple of high school kids who wanted a thrill, and that was the end of it.

So Michael and I went back home to Yonkers and continued to run WKOV and provide free radio service to our community.

Chapter Four
The Falling Star Radio Network

One day in May of 1970, a fellow who turned out to be a very good friend of mine heard WKOV. We were just playing music at the time. But he thought it was a little odd that anyone was broadcasting above the standard AM band. He listened a little longer, then he heard someone say, "This is WKOV — Your natural station on 1610 AM!" And that is when he knew he was listening to a pirate station.

A few nights later, he heard us again. He called up a friend who lived a little to the south and asked if he could pick it up. He tuned his radio to our frequency, and WKOV was coming in loud and clear. So he deduced, "Ah ha — they must be closer to him than me."

Using the old radio direction-finder technique, which is to take an AM radio and turn it different directions to see where the signal comes in the strongest, he decided to track us down. He grabbed a cheap old Lloyd's portable AM radio, got in his car, and went in search of WKOV. Eventually, he drove down Kneeland Ave. He found the spot where the signal seemed to be the loudest, got out of his car, and walked around the neighborhood.

When he came to our house, WKOV almost tore the speaker out of his radio. So he knew he had found it. About then, he saw some kids who were walking down the street and asked them who lived there.

A day or two later, I got this really mysterious phone call. I'd just come home from school and hadn't even had time to sign

WKOV on the air yet. Then the phone rang. I picked it up and a guy asked, "Is this Allan Weiner?"

He said his name was Joseph Paul Ferraro. Then he informed me that he had picked WKOV up on his radio and tracked the signal to my house. Now that was a pretty scary thing to hear — especially after what happened with WRAD. But then he told me that he was building a radio station of his own, and he really wanted to meet me.

He seemed to be genuinely interested. So I said, "OK, give me your address and I'll come over." I called my friend Kenny Sofer, who was 17 at the time and had a driver's license. When I told him about the phone call, he was very apprehensive about it. He thought it was some kind of setup. But finally, he agreed to take me.

After some searching, we found where Joseph Paul lived — an English Tudor house at 657 Warburton Avenue. I went up on the porch and rang the doorbell. His mom, Mary Ferraro, answered.

I said, "Hi, I'm Allan Weiner. Is Joseph Paul there?" She went into the kitchen and yelled down to the basement for him. (That's where J.P. was living. Everybody lived in the basement back then.)

In a few minutes, he came over to the door. And here was this guy with tremendously long hair, love beads, and a bandanna around his head. At the time, I was still a clean-cut American high school lad. I must have looked like a nerd to him.

I said, "Hi, I'm Allan Weiner." Then we shook hands and went down to his basement. Despite the contrast in our appearances, we had a lot in common right from the start. Joseph Paul had taken over the basement of his mom's house and set up a lab to work on electronic equipment. He also had a darkroom and an art studio down there. Joseph Paul showed me some of the transmitting equipment that he was attempting to build and told me how he felt about what was going on politically. Before our first meeting was over, I knew for sure that I wanted to get him involved in what we were doing.

I said, "I'll tell you what — I've got plenty of transmitters at home. I'll give you one so you can go on the air and we can become co-stations." And that was the start of something called the Falling Star Network — a group of free radio stations whose goals were to liberate radio, give people a voice, and stop the war in Vietnam.

Joseph Paul and I got into his 1953 Plymouth and drove to 245 Kneeland Avenue. I took him down to the basement and showed him two transmitters he could use with the least amount of modification — the old radiotelephone set we used with WRAD and another transmitter I'd bought for later use. I told him, "You'll have to build yourself some power supplies, but this will get you on the air quicker and with a lot more power than what you were putting together in your basement."

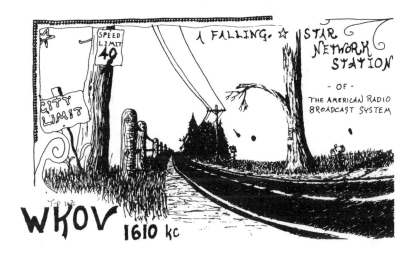

He took the old BC191 military transmitter home and went to work building the power supplies and setting up his station. Joseph Paul's father, who died when he was five years old, was into electronics. He was into building television sets and radios and stuff like that. So Joseph Paul had a basement full of parts, transformers, capacitors, and other great stuff to work with.

Joseph Paul had a lot of audio equipment on hand for his studio because he was into tape recorders, stereo systems, and things like that. He was attending Westchester Community College and he worked for the college radio station, WCC, which was really just a glorified PA system. Joseph Paul was able to get hold of WCC's surplus 8-channel audio console for his station. It was a pseudo-homebrew job with six or eight audio inputs. But still, it was a real improvement over what we were currently using.

Joseph Paul had to compromise on a lot of stuff. There was no audio processing, and the signal went from the power amplifier right into the audio input of the transmitter. It was kind of like a Rube Goldberg setup. But it worked, and it sounded very good on the air.

Next, Joseph Paul went to work on the antenna system. He erected a 20-foot wooden tower out of 1″ x 1″ lumber and mounted it on top of the house. And on the top of the tower, he clamped a big whip antenna.

We put the antenna up in just one day. We ran a transmission line down to the transmitter and ran some tests on 1620. The signal was great. His house overlooked the Hudson River and the soil acted as a great ground plane for the station. The BC191 was only capable of generating 50 to 75 watts of power, but the signal went far and wide.

Joseph Paul took to the airwaves in the spring of 1970. He came up with the call letters WFSR, standing for Falling Star Radio. We both used 1620 for our broadcasting frequency, so his broadcast schedule was staggered with mine. He was on the air Tuesdays, Thursdays, and Sundays, and WKOV's programs were aired on Mondays, Wednesdays, Fridays, and Saturdays.

With the new arrangement, at least one of our stations was on the air every day of the week. Sundays were Joseph Paul's favorite. Joseph Paul and I had a great time with our stations. Our efforts to keep WKOV and WFSR on the air turned into a decades-long friendship, and we always helped each other out in preparing programming and staffing people.

As the months went by, we received lots of letters from people in the community. And we knew that quite a few people from school were tuning in. But Joseph Paul kept telling me that we needed to go to FM. He said that's where all the alternative radio was being done — and to find the kind of listenership that we really wanted, we should go there.[1]

My friend, Albert Langella, took an old 50-watt AM VHF military transmitter and converted it to FM. It only tuned the upper end of the FM band, so we set it on a frequency around 108.1 MHz. Again, we didn't have any frequency counters back then, so the only way we knew where we were on the band was to listen on a receiver.

The next thing we had to do was build an antenna. I needed some tubular aluminum elements for the project, and my friend Larry Frumkies, said, "Why don't you go over to my apartment building and go up to the top floor? There are some unused TV antennas up there and you can probably yank some parts off." So that's what we did. Larry drove me over in his car and I climbed up on the top of the building and took the parts down. Looking back on it, it's a wonder somebody didn't call the police on us. But we were lucky, I guess.

After I finished building the turnstile antenna, we lashed it to the railing of the sun porch and ran a transmission line down to the basement where the transmitter was. Our house in Yonkers was at a

---

[1] Joseph Paul lived right down the street from Major Edwin Armstrong's former residence. Major Edwin Armstrong was our hero. He was the inventor of regeneration and the superhetrodine receiver, which is the basis of the design used in all radio receivers today. He had a laboratory up on the top floor of the house where he discovered regeneration, super regeneration, and did some of his pioneering work on FM broadcasting. So I guess it was just poetic justice that one of our stations, the station that Joseph Paul built, was right down the street from the house Edwin Armstrong was raised in.

relatively high location, so we were optimistic that the signal would get out well.

We decided to call our FM station WXMN. Major Armstrong's first experimental FM station in the country was known as W2XMN — the 2 denoted "experimental — New York State." So I said, "Let's just delete the 2 and call it WXMN." Those call letters weren't assigned to any other station. So, we went with that.

We put WXMN on the air — and what a difference! The reaction was instantaneous. We gave out the phone number to see if people were listening, and we soon discovered that our 50-watt transmitter covered the whole city of Yonkers and beyond. We got calls from listeners as far away as New Jersey and upstate New York.

I was really stunned. The response for one day's broadcast was much larger than an entire month on 1620 KHz. Joseph Paul was right — FM is where it's at. Of course, WFSR was still on the air as an AM station. But J.P. and I were looking into getting another transmitter so he could put a station on the FM band as well.

But one night in the winter of 1970, disaster struck. I was listening to Joseph Paul's program on WFSR at my house while I was modifying a 500-watt transmitter I intended to use to boost the power of WKOV. My head was deep inside the transmitter when I heard him say, "Due to circumstances beyond our control, we have to leave the air." Click.

It shocked me so much that I banged my head on the top of the transmitter. I thought, "Uh oh..." Then I grabbed the phone and called J.P.'s house. His mother answered the phone. I asked, "Mary, what's going on?"

She said, "Allan, the FCC is here."

That was it. I hung up the phone and called one of my friends to get transportation. We drove over there and sure enough, WFSR had been visited by the FCC. The agent had tracked down his signal, inspected his station, and informed Joseph Paul that WFSR was in violation of Section 301 of the Communications Act.

WKOV/WXMN studio, 1971.

In a few days, J.P. received what is known as a loop letter. It said, "You have been found in violation of Section 301... etc." And that was all that happened. Back then, the FCC just gave you a visit, sent you a warning letter, and that was the end of it. They didn't confiscate everything like they do today.

After the bust, everybody was pretty shook up for a while. We suspended all operations for a time while we decided what to do. As far as we could tell, WFSR was not ratted out or turned in to the government by anyone. I guess the FCC just picked up our signals at their monitoring stations throughout the country, triangulated it, and there we were.

There was a time when J.P. and I tried to do it legally. We didn't want any more trouble from the government, so we contacted the FCC down in New York City where they had a field office. We told them what we wanted to do, and they said, "Do you have a million dollars? Because if you don't, you're not going to get anything." Naturally, this was very disillusioning. We were a couple of young kids who really wanted to be on the air and do something beyond ham radio — and that was what our government told us.

After that we decided, "Hey, we have the same right as anyone else to use the public airwaves — the Constitution of the United States gives us the right of free speech, and we're going to use it." If we wanted to operate a radio station, our only alternative was to do it without the FCC's permission.

We didn't discount the fact that the government needs to regulate the broadcast bands to avoid total chaos. But we tried to work with the FCC, and they laughed in our faces. We were broadcasting on frequencies that didn't cause any interference and we weren't hurting anyone. How could the government claim that we were doing anything wrong? So about two weeks after the FCC's visit to WFSR, we both decided to go back on the air again.

But the FCC wasn't our only obstacle to staying on the air. The FM transmitter I used for WXMN had been modified to put out 50 to 75 watts, but it was really designed to put out only about 20 watts. So its power amplifier was constantly overheating. One day when the transmitter died, I discovered that the heat had literally melted one of the tuning capacitors. I fixed it and got it back on the air, but it generated so much heat that the tuning capacitors were going bad all the time. So I had to set up a huge blower that blew air into the transmitter to keep it from literally burning up. It worked that way for a couple of months.

Then one night when we were on the air, the transmitter caught fire and went out in a blaze of glory. Fortunately, it didn't burn anything else down — but WXMN was now transmitterless.

Digging up equipment had always been something Joseph Paul and I were experts at. We thought about building another transmitter, but decided to buy one instead. I knew that I could probably find used or military surplus equipment that could handle the power for a lot less money than what it would cost to try and build something from scratch.

I went to the Canal Street stores to see what I could discover. Then I came across a company called Liberty Electronics — of all things. I saw an ad in which they were selling radios and all kinds of electronic equipment. So Joseph Paul and I drove down to their store on lower Broadway to check them out. We looked through the place briefly and picked up a catalog. I took it home and thumbed through it all afternoon. This was a day or two after the WKOV transmitter burned up, so it didn't take long for me to notice that they had some compact 50-watt TRC1 FM transmitters for sale. The description said that they operated on 110 volts AC and gave out about 50 watts on any frequency between 70 to 100 MHz.

I thought, "Hmm, these sound just about right!" So I called Liberty Electronics and asked how much they cost. They quoted some outrageously high price like $100 or $200 each, and I said, "Wow! That's quite a bit of money. Can I come down and take a look at this stuff?"

We got in Joseph Paul's car and drove down to Liberty Electronics. The guy that owned the place took us into the back room where he had stacks and stacks of TRC1 transmitters. We told him what we were doing, and he thought it was a neat idea. Then I said, "We'd like to have one, but we can't afford the kind of money you're asking."

He looked at me and said, "Well, how much money do you have?"

And I said, "I don't know... I think I've got about a hundred dollars."

He thought it over for a minute and said, "I'll tell you what... give me a hundred dollars and you can take them ALL."

I said, "Take them all?"

"Yeah, give me a hundred bucks, but you've got to take them all."

So I gave him a hundred dollars and we ended up loading nine or ten TRC1 transmitters into Joseph Paul's car. We put them on the back seat, we put them on the front seat, we stuffed them in the trunk... by the time we finished packing his car, there were transmitters down to the wheel wells and I barely had room to sit. But we were both excited because now we had all the transmitters we would ever need. Not only could we get WXMN back on the air, but we could put WFSR on the FM band as well.

When we got back home, a lot of my friends were waiting for us. They knew that I was going down to New York to try and find some transmitting equipment, and were anxious to know how things went. They all helped us unload the car. Joseph Paul took two or three of the TRC1s home with him, and I kept the rest in the basement of my house.

Then we started to experiment on the transmitters and see what we had. Within a few hours, I was able to get one of the transmitters to operate. I had to rebuild my turnstile antenna, and J.P. immediately went to work constructing an antenna for his place. And just a few days after we got the transmitters, we had them up and going on 87.9.

We figured that if we had an AM and FM station, they should have separate call letters. J.P. wanted to choose a set of call letters that everyone would remember. He wanted the call letters of his FM service to stand out from all the other stations on the band. So he chose the call letters WSEX. Need I say more? Obviously, everyone got that one right. No one ever, EVER got those call letters wrong.

WSEX signed on the air and WXMN returned to the air with its new transmitter on 87.9 in December of 1970. Even though we moved to the other end of the FM broadcast band, the response was terrific. We had lots of people contacting us, lots of letters.

J.P. and I sat down at the kitchen table of my father's house one day to hammer out some kind of a name for our organization. I don't know why — I guess everybody just wanted us to do it. And

we came up with The American Radio Broadcasting System —
ARBS.

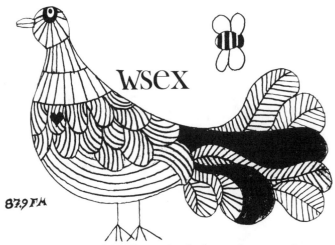

WSEX

87.9 FM

During the winter of 1971, we hooked up phone patches so we
could take telephone calls directly on the air. That was a new thing
for us. In fact, that was a new thing, period. Remember, at that
point, there were not any talk-radio programs per se. The only calls
that were taken over the air went through all types of screening and
seven-second delays.

Stations back then were very skittish about putting people
directly on the air. But airing live calls became one of our missions
in life. All of our stations — WSEX, WXMN, WKOV, and WFSR
— made a policy of taking live phone calls. We decided that one of
the main goals of the American Radio Broadcasting System, a
broadcasting service that now had four transmitters, was to give as
many people as possible free and uncensored access to the
airwaves. So we installed telephone couplers and telephone patches
and did it that way.

One day in January, 1971, the doorbell rang and I went upstairs
to answer it. I opened the door and said, "Hi, can I help you?"

And this guy said, "I'm from the FCC." Then he pulled out his
badge and said, "I'd like to see your station."

WSEX studio, 1971

So I said, "All right, come on in." Then he went down to the basement.

The agent took out his portable direction finder and started walking around with it. He asked me some questions, but I figured that the best thing to do was just keep quiet. (Of course, I had all the transmitters turned off at the time.) The basement was just filled with electronic equipment — not only transmitters, but radar equipment, antique radios, military communications sets, short-wave radios... everything you could imagine.

He walked up to a 1929 Atwater-Kent console table radio and asked me if that was the transmitter. Absolutely bewildered, I looked at him and said no. I was really shocked because FCC agents are generally radio engineers with degrees in engineering. He walked over to an old WWII radar indicator that I had on a table. And he asked me, "Is that the transmitter?" And I'm practically on the floor laughing. And I said, "No, that's not the transmitter."

Then he walked over to a transmitter that wasn't on the air and wasn't even hooked up and he asked me if that's the transmitter. And I said, "No, it's a transmitter, but it's not the transmitter we were using."

Finally, he walks over to the BC610 and he goes, "Is that the transmitter?"

And I said, "Yes, that's the transmitter."

Then he started copying down all the letters and numbers. He said, "Well, you're in violation of Section 301 and you really can't do what you're doing."

I argued with him and said that I can't get a license because the FCC won't let you have one unless you have a million dollars. Then I asked him who he thought we were harming. The agent said, "Well, you'll have to take it up with the Commission down in Washington because I'm just a field agent and I'm just doing my job." (Yes, and how many people have "done their job" for the government, resulting in so many senseless deaths?) He hung around for about ten minutes and then he left. About five minutes later, I noticed that he had left his radio direction finder behind. And it was still on.

I said to myself, "Oh good. I guess by default, I get a nice receiver." But in a few minutes, the agent came back to the house saying, "Gee, I forgot my RDF set." So he picked it up and left.

Four days later, I got another "loop letter" in the mail stating that we were in violation of this and that government regulation. So again, we went off the air for a while, had meetings, and pondered what we were going to do.

At one point, J.P. and I decided to send the following letter to the Federal Communications Commission:

> Sirs:
> You have asked us to write you an expatiation of on-the-air activities of the American Radio Broadcasting System (Falling Star Network). You have referred to us as illegal and unauthorized, and your representatives have spoken to us as if we were criminals for generating and modulating radio

frequency currents as if they were solid entities like telephones that we stole from you and hooked up to your lines without your permission. The distinction lies however in the fact that radio was discovered, not invented, and that these frequencies and principles were always in existence long before man was aware of them. Therefore, no one owns them. They are there as free as sunlight, which is a higher frequency form of the same energy.

Even you, the FCC, have stated that the airwaves belong to the people. You haven't lived up to it, because if the airwaves belonged to the people, we are the people. The airwaves belong to us. In FCC information bulletin 1B you state that any qualified citizen, firm or group may apply to the FCC for authority to construct a standard AM, frequency modulated FM, or television broadcast station.

Based upon this, we went about a year ago down to the FCC at 641 Washington Street in New York City to apply for a license. Our attempt proved quite humorous to your employees, who sent us away with words of "Forget it." Further investigations showed us why our attempt was then so comical. Licenses were so expensive and hard to get that even small stations were being sold for millions. Broadcasting was reserved for power men. A million dollars or so would get us a license and a channel pronto. The people, however, are not represented by millionaires. But that's who owns the stations. Various FCC agents who visited us to shut down the stations told us, "Get a first class license — for you it would be easy. You could then get a job at a station twiddling dials for someone else."

Most stations are designed only to make money. Music is sandwiched between commercials and ads, announcers tell us that programs will resume after station identification. That announcement is always followed by another commercial. But they are right, that's who they should identify with. Advertisers and money. WABC-AM in New York cuts a record by as much as one half so there is more time for ads. Is this radio for the people? American TV is among the worst in the world. And when you try to watch a good program, you are bombarded by three to five minutes of ads every ten. Is this for the people? You have set aside no bands for people to get started in broadcasting. Is this for the people?

While we do not claim to have the best programs on the air, we try to do our best. We don't claim to be the most free-thinking people around, but we give our microphones to anyone that has something to say on the air. We don't have thousands of dollars to spend on equipment, but we have taken our own money and made two high-fidelity stations. This is, we believe, for the people.

We are not disputing, however, your right to assign channels and set aside bands for the prevention of interference. We certainly, however, are disputing your right to reserve broadcasting for the well-to-do only. Our stations have had hundreds of people write to us. People who have written, "What kind of a wonderful station are you?" and the like. We have never had a letter of disapproval or complaint.

We started this whole thing because we love radio as an artistic and creative medium, and to bring freedom to the airwaves. Not because we want fat bank accounts and chauffeur-driven cars. We have chosen our operating frequencies especially so as not to cause interference with any other stations. However, as human beings and citizens of the United States and the world, we have a right to use the airwaves put there by whoever or whatever created the universe, and use them as we will. This is our freedom, this is our right.

Allan H. Weiner and Joseph Paul Ferraro of the American Radio Broadcasting System.

After a time, we signed our stations back on the air again. It was our only real option. We were operating on unused frequencies and weren't causing any interference, so how could it possibly be wrong?

It was one of our initial goals to build as many pirate stations as possible. We felt that the more stations there were, the greater the chance would be that the government would have to allocate frequencies for our type of operation. And as the old saying goes, if you've got one mouse on the floor it's easy to kill, but if you've got a whole room full of them, it's going to be a little harder.

Michael Schaitman decided that he wanted to put a station on the air. His parents had an apartment house down in Bronx, New York. So Michael set up a station in the corner of his bedroom there — WBRX — Bronx Radio. It consisted of a transmitter, a turntable, a tin-plated menace of an audio board, and an old Lafayette tape recorder.

But unfortunately, Michael had some problems with his equipment and WBRX didn't get out all that well. The transmitter did, however, modulate the florescent lamp in the kitchen, warm up his room on cold winter nights, and control the elevator in the apartment building. If anything, it proved that operating a station out of an apartment house is not the easiest thing to do. It can be done, but it's very difficult.

Another reason the station never got very far was that so much interest was focusing in on WKOV. We tried to become diverse and set up lots of stations, but WKOV always seemed to be the nerve center of our network. All the people who wanted to be involved with free radio were attracted to it. It seemed to go farther and have a more powerful signal than any of the other stations. So Michael eventually closed WBRX down and spent most of his time at WKOV, furthering the cause of free radio in Yonkers.

At one time, we counted about thirty to fifty people who were directly or indirectly involved with the radio station. One of the people who did a show once in a while was a fellow by the name of Andy Levitan. He was a really big guy with a gentle soul. Andy Lev-a-ton, as we used to affectionately call him.

The only problem I had with Andy was that he ate up all the food my father had packed into the refrigerator. And he drank up all the milk, too. I used to tell him, "Andy, you can just have one glass of milk and that's it." Otherwise, he would drink all our milk. I'll never forget the day that he came down into the basement with this quart-sized glass and said, "Well, Allan, it's one glass of milk!"

One night when I was having dinner, I heard an awful crash in the basement. I went downstairs and found out that Andy had dropped a gallon of wine which splashed all over our 300-watt AM transmitter, knocking it off the air. Needless to say, I was very upset about it. I cleaned off the transmitter and got WKOV back on the air as soon as possible.

In the spring of 1971 a friend of mine — David Haber, who was a very cosmic person — got interested in the station. He came to me one day and said that he wanted to do a morning show. I said, "Great!" He turned the station on the air at about six o'clock in the morning and played beautiful instrumental pieces. It's what people now call New Age music.

Michael Schaitman was my assistant engineer. He slept in the basement and kept an eye on things. Many times, the transmitter blew a fuse or one of its tubes would go soft. And Michael was there to pop in a new one and keep the station on the air.

Howie Schlifer also hung around the station a lot. The first time I met him was in the 8th grade. He ran with the rough-and-tough bully crowd, but he seemed to be a real intelligent guy and we got along well from the beginning. For a while, we lost contact with each other. And then during the radio-station period, we got connected again. He really liked what we were doing. He came over often to be on the air and have a good time.

A guy that went by the name of Brother Love also did some announcing for us. He drove around in a 1965 or 1966 Volkswagen, "the Beetle." He took the back seat out of it to make a place for his gigantic St. Bernard dog to ride. In fact, I think the St. Bernard was bigger than his Volkswagen. He took that dog with him everywhere he went. Brother Love heard us on the air one day, and he called us up and said, "Hey, man, I like what you're doing!" So we invited him to come over. Brother Love came down to the station every once in a while to talk about love and peace and getting it all together — all the wonderful stuff that people were talking about back then. And it was great.

Debbie and Lauri Strock, a couple of David Haber's friends, found out about the station in the spring of 1971. David told them he was working at this crazy unlicensed radio station, and they immediately wanted to become a part of it. The first thing they did was hand-print 25 big signs which they plastered all over the parks and public areas in Yonkers. They were sweet, wonderful gals and everybody loved them.

Our stations were very passionate. Many times people would get on the air and really cry their heart out about things that were going on — especially with all the needless killing and the war in Vietnam. And the radio station was a perfect vehicle for this. A lot of people wanted to say something, and here was a benign way to do it.

The standing rule at the American Radio Broadcasting System was that if somebody called up and wanted to be put on the air to talk about anything of significance, right after the next song it would happen. People also sent in public service announcements about a meeting or a demonstration and we ran those free of charge. There was no advertising whatsoever. We were very strongly against that. It was offered to us at times, but we said that we preferred cash donations or records or whatever you can give us. That way, we didn't owe anybody anything or have to feel obligated.

A guy named Peter was our mail drop. We had our listeners send mail to his home address. Then one day we realized that the promotional records and stuff that people were sending us weren't getting to the station. After that, we had our mail delivered to Jeffrey's address for a while. I think in the end, we just decided to have it delivered to my father's house.

You have to remember that back then, we were pretty damn blatant about it. The telephone number we gave over the air was the telephone number for my father's house. We just shut the phones off upstairs and patched the calls directly over the air. It was neat.

This was a new thing. In the early 1970s, it was unheard of to put people on the air without screening or without a seven-second delay. Nobody ever did it. But we believed that people should have free access to the airwaves. We did ask people to refrain from using vulgar language. An occasional word slipped through, but very rarely did anyone curse on the air. Taking phone calls from listeners was one of our mainstays. Even to this day, I'm very proud of it. We got a lot of heavy calls from anti-war protesters, draft dodgers, and Students for a Democratic Society-type groups. Our network was a real unifying resource — which is one of the nice things that small, low-power community radio services can do.

But for some strange reason that I never understood, my sister Barbara didn't mix with it at all. Barbara is only five years older than me and is really part of the same generation. It would have been very easy for her to come down to the radio station, see what was going on, and be a part of it. I always wanted her to do a program, or at least intermix with all the excitement and positive energy. But my sister never wanted anything to do with it. She was very detached about it all — which even to this day saddens me.

The Falling Star Network had staff meetings every week or two. They turned into a really nice get-together. We usually had about thirty people or so show up — anybody that was connected with the stations was invited. Sometimes, we announced them on the air so our listeners could come down. It was wonderful. We talked about the programming, the music, issues of the day, and so on. Everybody ate together, talked about the station, and did things like that. There was a lot of good harmony.

I had a red organizer book with all the days of the week listed in it. If somebody wanted to host a program, I said, "Okay, when?" Then I'd look through the book until I found a free time period that fit in with their schedule. It worked out very well, and we usually had more than enough people to fill our airtime.

Even though the Falling Star Network was a free radio service, there were some rules. You could not be drunk or stoned and be on the air. That was absolute. I think I only had to throw one or two people out of the studio because they just were a little bit too under the influence of one thing or another to be broadcasting. At that time, I didn't drink or smoke — period. Zippo. Everybody else did, but I had this strict thing about mind-altering stuff. (To this day, I don't smoke, do drugs, and rarely drink.)

All kinds of people were interested in the Falling Star Network. We even had a couple of guys who seemed to be from the Mob pay the station a visit. I'm serious! One day a couple of real mean-looking guys walked in the back door of the station. I asked them who they were, but they didn't say anything specific. All they said was that they were told to check us out and see what was going on. They looked around, and they seemed to really like it. They said, "We think what you're doing is okay. We know you're having a lot of trouble with the government and you're probably going to get more trouble from them. But we got no problem with ya — we think what you're doing is all right. Good luck." And that was it.

Those two guys that visited us were a riot. Whether they were from the Mob or not, who knows? But they kind of suggested that they were from the Mob — and knowing Yonkers politics, I wouldn't be surprised one bit.

The radio station was literally open 24 hours a day. People came in and out of the back door at all hours. At the height of it, just about everyone involved with the station was being tailed by the FBI and was under surveillance. I don't know why, because it was pretty obvious what we were doing. All you had to do was turn on the radio.

Words cannot convey the fun we had. Out of all the radio I have ever done, those free stations we operated in the spring and summer of 1971 were the best time I ever had. It was just great. The Falling Star Network was always the center of energy and excitement. People went to anti-war demonstrations with their friends, then they came down to the station to talk about what was going on. We always took lots of telephone calls, both on and off the air. Doctors, lawyers, and people from all walks of life called up or came down to the station to share their experiences with our listeners.

I guess that's why we were all being followed by the FBI. I know that I was tailed for weeks. And so were Larry, Michael, and everybody else. At one point, Michael and Larry Frumkies got on Michael's BMW motorcycle and led them on a wild-goose chase that ended up at the Croton Point dump. Michael's motorcycle got through fine, but the FBI car got stuck in the garbage. Michael circled around them a few times just for fun, then came back to the station.

But when the word got out around the neighborhood that we were being tailed by the FBI, some of my friends' parents got pretty upset about it. Kenny Sofer and Larry Rand weren't allowed to visit me. For a while, their parents forbade them to go anywhere near me because they were petrified that their children would be followed by the FBI as well.

Looking back on it, I can't really blame them. No parent wants their kids being followed by the FBI. My father, who is an attorney, looked into the situation, but the only thing he could find out was that special detectives from the Westchester County Police were on some kind of a confidential assignment. In reality, we didn't know who the hell it was. But it went on for a very long time.

In July, 1971, we were hearing a lot of rumors that the stations were going to be busted. It made us pretty nervous, especially with all these FBI and Westchester County detectives running around following everybody. So J.P. and I had a meeting and said, "Look, why don't we just go down to the FCC and talk with them and see if we can get some help." They had this radical commissioner by the name of Nicholas Johnson, who was kind of a free-thinking individual. He was written up in the *Village Voice* and in *Rolling Stone* magazine as an open-minded FCC commissioner who was going to help change the face of broadcasting. So we called up his office and made an appointment to see him.

We shut the stations down temporarily while we went down to see if he could do anything for us. When we arrived at the FCC office in Washington, DC, we met a whole bevy of people. We spoke to a couple of engineers and discussed with them the frequencies that we were operating on — which were 87.9 and 1620 AM. They said there had been some thought of allocating 87.9 for certain non-commercial use. But at the moment, it was just an idea. We spent quite a bit of time with the people in the engineering offices trying to convince them that we were doing was a good thing and they should leave us alone.

Finally, we got to meet with Nicholas Johnson. We were in his office for quite a while, talking with him about all kinds of stuff. He was very sympathetic. He said something like, "I really understand what you're doing, and I think what you're doing is neat, but there's nothing I can do to help you." In other words, he had an appreciation for what we were trying to do, he understood our plight, but under the present regulations, there was nothing he could do.

What we were really asking was for the Commission to just leave us alone — or else issue us an experimental permit, which would allow us to continue to do what we were doing. After that, we went to another engineer's office and discussed basically the same thing. They seemed sympathetic, but they really didn't want to get involved. They did not want to sanction our stations or grant us permission to broadcast.

When we left the office, we were frustrated. But in some ways, we were more committed than ever, because we realized that they — being the Commission and the people we spoke to — seemed sympathetic to the things we were doing, but they didn't want to do anything to help us. This, we felt, was not what was supposed to happen in America. Here we were, trying to provide Yonkers, NY, a city of 200,000 souls, with its first real radio station. We would have taken anything they gave us. We would have taken five watts, ten watts — whatever they let us have. But they gave us nothing.

We went back home, mulled it over and said, "It's hopeless to try and work with the FCC." Then we went back on the air.

Through the later part of the summer, the Falling Star Network was a 24-hour-a-day presence. The morning show with David Haber came on, and there would be a cast of thousands. Everybody went on the air, poured their heart out, and got very emotional about the war in Vietnam. They talked about the demonstrations, civil rights, individual rights, and things like that.

A lot of passion was in the stations. I can remember many summer nights when two or three people sat in the control room broadcasting commentary about the latest war crimes or deaths in Vietnam — the latest government atrocities and the military-industrial complex insanity. We took calls from people all over the city. Sometimes, people came down to the station to meet us and go on the air. It was just wonderful to be a part of it.

There was a lot of peace, love and understanding, let me tell you. There were a lot of girls and guys running around together and everybody had the hots for everybody else. But nobody ever got hurt and, thank God, nobody ever got pregnant. It was the last of an era. There was some pot smoking done in the basement from time to time. And in a few instances, I saw a couple of people underneath the rugs together when I came down to check on the transmitter in the morning. But I never said anything about it, because if they weren't doing it there, I knew they'd be out doing the same thing somewhere else.

Even though my father was aware of what we were doing with the radio stations, he stayed away from the basement. He never came storming down the stairs demanding that we shut the thing down, like a lot of fathers might do. But he was worried about what might happen to me if the government returned and made more raids on the station.

Actually, he was concerned about it to the point where he did things I didn't even know about until much later. As an attorney, he contacted his congressman and senator to see if there was some way they could have the FCC grant us permission to broadcast — or at least leave us alone. Unfortunately, nothing was ever done.

I know that he felt the war in Vietnam was wrong. He fully understood President Eisenhower's warnings about the military-industrial complex. My father is a very well-read and well-informed citizen. I've always been able to sit down and talk with him about anything political. And he's a compassionate, feeling man. He was in World War II. He saw the blood and the guts of war and felt it was wrong for the United States government to send American kids to get killed in Vietnam.

So undoubtedly he understood our reasons for operating the radio stations. True, it wasn't legal. But as an attorney and as a man of law, I think he viewed what we were doing as civil disobedience. In other words, we were breaking the letter of the law, but certainly not the spirit of its intent. We were just a bunch of kids who instead of being on the street and getting in trouble and getting into drugs, were down in the basement running a radio station and talking to our audience about things like "stop the war" and "the peaceful, non-violent way is the only way."

J.P.'s mom, Mary, was even more opinionated than my father. She was absolutely 100% against the war. She did not like President Nixon, and she had a very, very strong distaste for what the government was doing to the kids. She had a full understanding of what the young generation was trying to say. Of course, she was concerned that her son, J.P., would get hurt by the government or get arrested. But she definitely understood what we were trying to do and she did not attempt to stop us. Like my father, I think she considered us to be a group of people trying to say what we had to say with no other venue to express it except through civil disobedience.

Remember, back in the early 1970s, you couldn't buy time on a radio station. There was no way to get on the air except for a brief interview program, and even they were very controlled. You couldn't express your views the way you can today.

Life at my house was crazy, with so many people coming and going. There was never a dull moment. But there seems to be this unwritten rule in life that all good things must come to an end. And the Falling Star Network of the American Radio Broadcasting System was about to be destroyed in one fatal swoop.

A friend of mine who belonged to the Yonkers Amateur Radio Club, Michael Kusiac, called me up one day and said that the head of the Yonkers Amateur Radio Club — a guy named Otto — was on a one-man crusade to destroy us. For some weird reason, Otto thought we were communists who were out to destroy the United States of America.

I said, "My God, this guy sounds crazy!"

Michael told me that everybody else in the organization was telling Otto to just leave us alone. We weren't hurting anybody, we weren't on the ham radio bands, we weren't interfering with anybody, we put out a clean broadcast signal, so why bother? But despite all that, he continued his one-man crusade to get us off the air.

The FM and TV transmitting antenna
on the roof of my Dad's house, 1971.

At one point, he even called the station and asked us why we didn't get a license. I said, "We went down to the Commission and there are no licenses to get." He asked why we didn't become amateur radio operators and I told him, "If we become ham radio operators, it stipulates directly in the rules that you can not broadcast — and we are broadcasters. We want to talk to a wide group of people in the community, not just do point-to-point stuff." And that was the end of that discussion.

Otto continued on with his crusade and put lots and lots of pressure on the Federal Communications Commission. At the time, according to the Yonkers Amateur Radio Club, the FCC didn't really intend to do anything about us. We were not causing any interference, and even though what we were doing wasn't legal, we were a low priority in their book. But Otto refused to give up. Apparently, he put pressure on the District Attorney's office — who in turn put pressure on the FCC and forced them to do something.

August 11, 1971, was the last official day of the Falling Star Network. We had been hearing a lot of rumors about another government raid on the stations and we were contemplating going off the air for a while just to let things cool down. But it was too late — the wheels had already been set in motion for the destruction of the Falling Star Network.

At 8 o'clock in the morning on August 12, 1971, I was sound asleep in my bed. Then all of a sudden, I felt a cold hand grab my arm and shake it. Then an unfamiliar voice said, "Allan, get up." When I opened my eyes, I saw this guy with black hair, a flat-looking face, and a dark suit shaking my arm and demanding that I get out of bed.

I looked at him, wondering if I was having a nightmare and said, "Who are you?"

And he said, "I'm a federal marshall, and you're under arrest for illegal broadcasting. Get up and put your clothes on."

My father was standing near the door. He said, "Son, you'd better just do what they say." So I got up and I put my clothes on. I didn't say a word. They escorted me downstairs to the basement. There must have been a half dozen people there. I looked out the window and saw a huge moving van parked in front of the house. And men from the government were already starting to haul our equipment out of the basement.

I didn't say anything. Somebody read me my rights and told me that I was under arrest for violating Section 301 of the Communications Act. They asked if there were any bombs, guns or explosives in the house, and I told them there weren't.

When I was allowed to use the telephone, I called J.P. His mother, Mary answered the phone. I said, "Mary, the FCC is here."

And she said, "Well, they're here, too." It was a simultaneous raid on WXMN, WKOV, WSEX, and WFSR.

For about four or five hours, I was held in detention. I wasn't even allowed to go to the bathroom without an escort. They had me in handcuffs, leading me around the house while they totally destroyed the station. They tore down the antennas, cut the transmission lines, and ripped out all the power cables. They took everything — transmitters, turntables, tape recorders, microphones, records.

I begged them not to take anyone's personal stuff and they said, "All right — but you've got to point out what it is."

So I went down to the basement and said, "Well, that tape recorder belongs to Kenny, all the records belong to Larry, and this belongs to this person and that belongs to that person." And they said fine.

I took full responsibility. I told them, "Nobody is involved with this station except me."

And they said, "Well, we're not interested in anybody else but you."

They asked me a few more questions, but I didn't really respond to them. After about three or four hours of this nonsense, they put me in a car and drove over to Warburton Ave. About half an hour later, they escorted J.P. out of the house and put him in the car with me. They handcuffed both of us together. And I kept thinking, "I can't believe this is happening just because we were running a radio station!"

We were brought down to Foley Square, photographed, fingerprinted, and thrown into a cell. And there we were with all the mother rapers and robbers and drug dealers. Just like in *Alice's Restaurant*, people came up to us and said, "What are you in for, Mac?"

And we'd say, "Running a radio station."

Someone said, "Running a radio station? What kind of a radio station?"

And I said, "You know — an AM and FM radio station — only we didn't have a license."

One guy said, "You got arrested for THAT?" I mean, broadcasting without a license was nothing to these people. They were arrested for attempted murder, assault and battery, and heavy stuff like that.

After we had been in the cell for a few hours, my father came down. I could hear him in an adjacent office telling the District Attorney, "My son took all this equipment and put a radio station on the air and tried to help people! Treating him like this is atrocious!"

We were arraigned in Federal District Court. The magistrate was very nice about it. He said, "Look, I'll release you on your own recognizance (which means no bail), just promise not to go on the air until this case is disposed of."

And we said, "Fine." Because we knew it was over.

Our case got a lot of media attention. The press was in the room when we were being arraigned. We were interviewed by all the newspapers and the media. In the middle of all this, my father came in and offered to take us home. But since we were still being interviewed by the media and had a chance to present our side to the public, we hung around for a while.

When I got home, I walked down into the radio station and it was just a gutted husk. They only thing that remained was the console table. Everything was gone — right down to the blasted extension cords. It was ALL gone, even the program books and the logs that we kept of the operating perimeters of the transmitters. I felt totally raped and exhausted.

The phones rang off the hook all day. People wanted us to come down and do interviews on all kinds of talk shows, so I guess that was some consolation. At least they had an interest in our situation. There's always more interest from the media when they start putting handcuffs on you and throwing you in jail. Overall, the media was very good to us. They were sympathetic and treated us fairly.

Needless to say, a lot of the people that used to hang around at the radio station scurried away like ants that were afraid of being stepped on after the bust. I remember somebody saying, "Well Allan, you're going to find out who your real friends are now." And for a while, everybody was spooked. I think it took a week before Larry Frumkies and the rest of the people who helped me run the station even dared to come near the house. But when enough time had passed and the memories of that horrible August day started to fade, people began to come back. And everybody wondered what we were going to do.

In those days between the bust on August 12 and the sentencing, there were lots of radio interviews. We were interviewed on WBAI, and all the local college stations and FM stations. We were interviewed on Channel 5 in New York. And we got lots of phone calls, because more people than ever knew about our situation and wanted to know more about what we were trying to do. We were still trying to think of a way to broadcast legally, but every option we examined required a lot of money — which we didn't have.

The later part of the summer was spent in high anxiety about the October court date. J.P. and I kept wondering how bad it was going to be. At one point our attorney said, "Well, the judge may require you to go into rehabilitation."

I said, "Rehabilitation — what are they going to do? Start us out with a broadcast transmitter, and then wean us down to a ham transmitter, and then wean us down to a CB radio, and then wean us down to a Lafayette Broadcaster that only goes 50 feet, and then down to a wireless microphone? Are they going to send us to a radio rehabilitation farm?"

J.P. Ferraro in front of the Edwin Armstrong Laboratory, 1971.

My father hired an attorney to represent us, but we were still pretty nervous about the whole thing. Free radio is one of those victimless crimes — who could the judge possibly claim we hurt? But still, there was talk of a possible jail sentence, fines, and all that. So we really had no idea of what was going to happen. It all depended on the judge, we were told.

Finally, the day arrived. We went to court, stood before the Federal Judge, and pled guilty to the charges. And boy, did that judge give us a tongue-lashing. He glared at us and said, "It's Section 301 of the Communications Act, and you can't violate the law. You both should know better." He went on and on about it. Then he told us we were both sentenced to one year of probation.

So that was it. Many people thought that a year's probation seemed to be a bit too much. But the government wanted to make sure that we stayed off the air, and I guess they felt they had to do something. We were supposed to be treated under the Youthful Offender Act, which meant that young people can be given a sealed file so it doesn't follow you for the rest of your life. But unfortunately, it never happened, and the charges against us were put on public record.

Chapter Five
The Move to Maine

Even before the FCC busted our radio stations on that fateful day of August 12, 1971, we were looking for a house — a place where we could all live and run our radio stations. 1971 was a good time to buy property. Houses in the New York area could be had for as little as $10,000 back then.

But after the government raid, I was fed up with it all and I wanted to move someplace far away from New York. One day I was telling J.P. how I felt and he pulled out a map of Maine and said, *"This* is where you want to go." At the time, I hardly knew anything about Maine. But if J.P. liked the place, I thought it was worth looking into.

My friend David Haber was attending Ricker College, a small school that had an enrollment of about 200 students up in Houlton, Maine. So I told J.P., "We can go visit David and use his dorm room for a base of operation."

We got into J.P.'s car and drove up to Maine in late August, 1971. This was the first time I had been on a long road trip. We drove and we drove and we drove and we drove. Something like twelve hours later, we finally reached Houlton. By then, it was night, and we crashed (now that's an old term!) in David's dorm room.

I liked the town right from the start. Houlton is a small farming and lumbering community of about 9,000 people, with a lot of nice quaint buildings. And I was very impressed with how friendly people were in Houlton. They'd say, "Hello," when you walked down the street and everybody had a smile on their face.

We were pleased to discover that land was incredibily cheap in northern Maine. Back in 1971, you could buy hundreds of acres for very little money. So we decided to look around and see if we could find a suitable place to buy. My mom had left me some money when she passed away in 1970, so I had that to work with.

We went to a broker and started looking at farms, tracts of land, houses, buildings — all kinds of property. Finally, the broker showed us a 100-acre potato farm up in Monticello. The property was beautiful. There were fields, a stream running through the land, nice woodlands, cedar swamps, all kinds of different topography. There was an old farm house, a couple of garages, a huge barn, and even some farm machinery. I told my father that the farm looked like a good investment, and he agreed. We negotiated a price of $12,000 for the property — that was for 100 acres of land, all the buildings, plus 350 potato barrels and two potato harvesters, one tractor, and two farm trucks.

As soon as I purchased the farm (I think the papers were signed in October of 1971) I started making plans to move. I had a lot of stuff to transport up there. So we rented the biggest U-Haul truck available. All my friends pitched in and helped load up the truck. By the time we finished loading — which took about three days — the truck was packed solid. I had parts, electronic gear, clothing, chairs, and tables. Then my father gave me a whole bunch of furniture, silverware, and household-type stuff, so I'd have something to furnish the house with once we got there.

Everyone was sorry that I was going to move away, but I had really had it. After being arrested and all that stuff, I was fed up with the rat race. So that night at 9 PM, J.P. Ferraro, Michael Schaitman, my cat Dudley and I climbed into the truck and began the journey up to Maine. We were all tired. But we were anxious to leave and wanted to get the hell out of New York. We got on Interstate 84 and drove all night. J.P. and Michael were the only ones who had licenses, so they took turns driving.

Of course, there had to be an incident. And it was a pretty bad incident, too. We were about 130 miles away from our destination, a little south of Bangor, Maine. At about 5:30 in the morning, Michael was driving and he fell asleep at the wheel.

All I can remember is the truck rolling off into a ditch, hitting a rock embankment, and the box of the truck literally being ripped open like a can of sardines. Equipment, clothing, furniture, transmitters, power supplies, and television sets were strewn all over the place. It was a spectacular crash. And miracle of miracles, every part of that truck was totally destroyed — with the exception of us. Somehow the cab escaped being crushed. J.P. and Michael crawled out of the cab without any damage. I got a broken wrist in the incident. And Dudley the cat was fine.

Needless to say, the wreck was a mess. There was stuff strewn all over the highway and down in the ditch. I had to go to the hospital to have my arm fixed. Michael stayed at the scene and hired a couple of trucks. They picked up as much as they could that hadn't been destroyed, and it was all delivered to the farm in Monticello later on that day.

We arrived at the farm on November 7, 1971. It was a great start to my new life in Maine — with a broken wrist and all of my worldly possessions scattered over the highway and destroyed. But the important thing was that nobody was injured too badly in the wreck. And looking back on it, most of the stuff I lost was junk anyway.

That winter, I didn't do any broadcasting whatsoever. After being busted and all that crap a few months earlier, it just wasn't something that I was going to do. J.P. stayed at the farm house for about two weeks, then decided to move back home to Yonkers and get a job. Michael hung around with me for a couple of months, then went back to Canada to visit his parents. After that, he took a job in a furniture factory. All the people who were going to move up to Maine with me decided not to do it. And I was there all by my lonesome.

Eventually, David Haber decided to move in with me, but he was the only one. So I decided to take in boarders to help with the living expenses. At least I didn't have a mortgage, because the place was bought and paid for with cash. And the taxes were only about $80 a year.

*Allan H. Weiner and David Haber at the farm
in Monticello, Maine, 1972.*

It was a long winter. I had never experienced a winter such as that before. I basically stayed at the farm for the whole winter by myself in quiet solitude. I read a lot, built a lot of electronic devices, played around with television equipment, and just sat around trying to figure out what to do with my life after the "radio daze" of six months earlier.

In the spring of 1972, I met my wife-to-be Sarah. She was going to Ricker College and had a dorm room there. We didn't really date — it was one of those love-at-first-sight things. We were kind of interested in each other from the first time we met. Sarah came out to my house on a visit with some friends on April 3, 1972. We met, immediately got together, and had a great time. We had a beautiful beginning.

Later on in the spring of 1972, I decided that it might be a good time to go to college. So I enrolled in Ricker College for the September session. While going to school, I stayed with Sarah in her dormitory room, which she had maintained. It was my first year and her second year in college.

Since I arrived at the farm, the closest thing I did to broadcasting was to hook up a few 100-watt AM transmitters to run some short tests. Up in northern Maine, there really weren't that many places to broadcast to unless you ran a lot of power. If you take a map out, you'll see that Monticello, which is 15 miles north of Houlton, Maine, is in the middle of nowhere.

So the first thing I thought of doing as soon as I got to the college was to build a radio station. I designed a carrier current station to broadcast to the college area on 1600 AM. Radio WRL was housed in Sarah's dormitory room, because there was really no other place for it. I built a console, modified an old surplus transmitter, and hooked up a capacitive coupling unit so I could feed about 10 watts of power into the college's power system.

Our station was an immediate success. WRL was so popular that the next year, the Ricker College Student Senate invited us to move it to the Student Senate Building — an old town house that was used for student activities.

Then a few people came to me and said, "Why don't you try to get a non-commercial educational license for Ricker College and go on the FM band?" Before I knew it, everyone, including the school officials, was asking me to apply. So I got the necessary forms and documents and started the process of applying for a non-commercial 10-watt educational FM station.

After a couple of months of filling out forms, doing the surveys, taking the photographs, and the whole bit necessary for a construction permit, I sent in an application, which was signed by the Board of Trustees of the college. Ninety days later, we were notified by telegram that the FCC had granted us the permit. Our call letters were WRNE.

The student senate allocated about six or seven thousand dollars to build the station. We bought all-new equipment. I got a Gates 10-watt FM transmitter, a small 2-bay transmitting ring antenna, a brand new audio console, a couple of tape recorders, two new phonographs, and a record library. It was a first-rate college station.

Sarah LeClair at the controls of WRNE,
Ricker College, 1974.

But shortly after I put WRNE into operation, the station I designed and built was taken away from me. The student senate got together and voted in another station manager. I guess they thought I was too radical because my hair was too long for them or something. All they wanted me to do was be the chief engineer. Everybody except the student senate wanted me to run the station. But I really didn't care that much because I still got to do the engineering and keep the thing on the air. And that in itself was a lot of fun.

Overall, we did very well with WRNE. It went on the air as the second FM radio station for the Houlton, Maine, area. It was on for about twelve hours a day, and was staffed by the students on a voluntary basis.

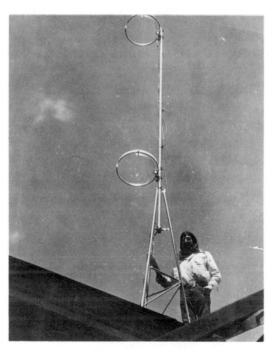

Allan H. Weiner with FM transmitting antenna for
Ricker College station WRNE, 1974.

WRNE was pretty much a free-form station. There were all kinds of music, interviews, talk shows, commentary, and so on. When the communications department was built a year later, WRNE was integrated into various class programs. All kinds of crazy things happened there. College radio is weird.

But then, the student council decided to appoint a media board to ensure that the people that were running the radio station were responsible. Now a media board is a body of people who are supposed to oversee the operation of the radio station and make sure that nobody says "shit" or something like that. I didn't see any real reason for having a media board because it is just another form of censorship. I was steadfastly against it. But the college wanted it, and this idiotic body of fellow students insisted on having it. So a media board was put in charge of WRNE's operation.

Sarah LeClair & J.P. Ferraro at 1st Annual
Pirate Radio Conference, 1972.

It was then I realized all of the awareness gained in the later part of the 1960s was slowly eroding away. I saw my so-called friends, people that I knew, fellow students who seemed to be just as open-minded and radically free-thinking as the next person, get appointed to this "media board" and turn into a bunch of fascists. They were more than happy to limit people from doing their programs because they were putting too much controversial material over the air and somebody might not like it.

The media board reigned over WRNE for two years. Then finally, it was dissolved over an incident at the station. A friend of mine said something they didn't like on the air, and they cracked down on him. I came to his defense and entered a resolution that the media board was a totally worthless entity that should be disbanded. Believe it or not, it passed — and no more media board!

Later on, I put a cable TV station on the air for Ricker College. The local cable television system in the town gave us a channel on which to air our programs. Houlton had a small, 13-channel cable system, and Channel 12 wasn't being used at the time. So they gave it to Ricker College to use for the benefit of the local community. We called it Channel 12 Television.

We built a small TV studio in the basement of one of the dormitory buildings. The local television stations donated all kinds of old TV and studio equipment to us. We had a lot of fun with the station. We aired news, movies, and all kinds of material.

Ricker College had some really fine instructors. One of the instructors that I made a very close friendship with was a gentleman by the name of Dr. Shakir. He was the director of the school's Department of Muslim World Studies. Ricker College was one of the only colleges in the United States that had a program in Muslim world studies, which was funded by a grant from Saudi Arabia. Dr. Shakir was an amazing man. He could speak four or five languages. He was incredibly intelligent and was versed in just about everything. You could go to him any time day or night for either an academic problem or a personal one.

He taught me a lot, influenced my life a lot, and generally raised my awareness and my consciousness. He was not only the Director of Muslim World Studies, but the Director of the Political Science Department as well — which was a pretty big department at the college. He taught me more about the political process and the essence of the human being than anyone else.

I remember one day I was lamenting to him about how we just seem to be destroying ourselves — human beings just can't get along, the political systems are so corrupt, and everything is such a mess. He said that change happens very, very slowly because the level of awareness of a human being takes a long time to raise. The only time that you get an extremely significant change in any type of a social structure, he said, is when the level of awareness of human beings raises up a step.

We would talk about these things for hours, Dr. Shakir, Sarah, and I. Dr. Shakir was a guiding force, almost my fatherly protector when we were living up in Maine.

While I was attending Ricker College, I was hired as the chief engineer of the school. I was put in charge of the radio and TV stations, as well as all the other electrical stuff that needed to be taken care of — PA systems and things like that. And as soon as they hired me as a faculty member, I didn't have to pay tuition anymore.

So after two years in Maine, I was working at the college and attending college for free. Then the school asked me if I'd like to teach a couple of courses. I developed a curriculum for radio broadcasting and broadcast history. I also taught a couple of classes in radio electronics and one class in radio engineering. I always got a lot of satisfaction out of saying that I received a degree in communications at Ricker College from a department that I created. Some of the best years of my life were spent at that college.

While we were still going to school, my wife and I decided to diversify a little bit. We opened up a natural food store in the downtown Houlton area. We called it "Uncle Fred's Natural

Foods." It was right in the middle of downtown Houlton. It was a great store. We sold everything! We even made our own peanut butter and had a flour mill in the back. One time a friend of mine was pouring honey for a customer who wanted to know how fresh it was. And he said, "Why this honey is so fresh, I wouldn't be surprised if a bee flew out of this can right now!"

In the 1970s, a lot of small New England colleges were lapsing into insolvency and closing their doors. And Ricker College was going downhill as well. The college hired an administrative staff, namely the president, vice president, dean of instruction, and a few other people, who were literally a bunch of crooks. It was horrible to see the kind of charlatans and jerks and just plain assholes Ricker College employed to try and save the school. They took money out of the general fund and used it to remodel their houses, repave their driveways, and things like that. Needless to say, the school's enrollment kept going down and the debts kept rising.

It was another major case of being disillusioned. After spending so many wonderful years at Ricker, the decline was a very painful process to me. The corruption at the school was just disgusting. Dr. Shakir, myself, and a group of other faculty members knew this, and we tried to do what we could to change things. I remember the many screamathons we had with the Board of Trustees when they held their meetings. But it did no good because the Board of Trustees didn't really want to know what was going on.

The college eventually had to close its doors in 1978. It was a tremendous loss to me personally, because all of the work and effort I put into building the communications department and developing the radio and TV stations for the school went down the drain. It was also a real big blow to the town of Houlton, Maine.

I did get to graduate, by the way. But I was so busy running the equipment that I didn't even get to go up on stage to get my diploma. I graduated with a Bachelor of Science Degree with a Major in Communications. It's nice to graduate from a department that you founded. Not very many people can say that! I don't mean to brag — but that's what happened.

About a year before the school closed, I'd decided to apply for a license for a small non-commercial 10-watt FM station at my house in Monticello, Maine. I had to form a non-profit organization, which was real easy to do. I sent the application to the FCC and in three months, it was granted. I was even able to get funds from the federal government to hire a few people to help me run the operation.

I built the studio in a room above my kitchen. I bought an old Gates 10-watt FM transmitter and hooked it up to a 2-bay ring antenna. Amazingly, I was able to resurrect the call letters WXMN, which I had used for my pirate station in the Yonkers back in the early '70s. The station was licensed to the Monticello Community Broadcasting Company. WXMN broadcast on 89.5 FM. The signal went about 15 miles and it covered the town of Monticello. It didn't have that much of a distance, but it was your real down-home local station in the area.

We were on the air about twelve hours a day, mostly with the help of volunteers. We had all sorts of programs and played every kind of music imaginable. We even had people coming down from Canada to produce news programs and things like that. Monticello, a town of about 1,000 people, had an annual town meeting. We'd take our tape recorders down, record the town meeting, and play it back over the air so that anyone that wasn't able to attend could hear what went on.

At that time, I was getting unemployment compensation from the college fund, which supported me and the radio station. My wife Sarah was working and we still had the natural food store. Sarah and I had a printing press in the basement of the store. We printed up invitations, business cards, whatever anybody wanted.

So for a while, I was doing a lot of things. But the money still wasn't quite what it should be. I was entertaining the thought of building a commercial FM or AM station at my place in Monticello. But in the meantime I had to eat, so I started looking for a job.

In the spring of 1978, I went to New York City and took the test to get my first class radio-telephone permit, which was required at the time in order to work as an engineer at a radio or television station. I passed it, and then began looking for a job in northern Maine.

I called Malcolm York, an engineer who did some work for the local radio station in Houlton, Maine, and asked if he knew of any work around the area. He said, "Well, there is a station in Presque Isle, Maine — WEGP — that's looking for an engineer. They might be interested in hiring you."

So I called up WEGP, got the manager on the phone, and said, "Hi, I hear you're looking for an engineer — that's me!"

He said, "Yes, we are — why don't you come up and we can talk about it?"

So the next day I hopped in my car, a 2-cycle Saab, and drove to WEGP. The station was located on Main Street in Presque Isle. Gerry Pratt and Harold Glidden[1] interviewed me. They asked me what experience I'd had. And I said, "My heavens, I built this radio station and I built that radio station..." (True, they were pirate stations, but I left those details out.) "I've built transmitters, I've built receivers, the whole bit."

They walked out of the room for a few minutes. When they came back they asked, "Well, how much money would you like?"

I said, "I don't know... how does $250 a week sound?" Now mind you, this was back in 1978. At the time, I would have settled for $150 a week. They went outside and mulled it around and came back in and said, "Well, how's $225 a week?"

I said, "Fine, no problem."

Aside from the job at Ricker College, it had to be the easiest time I've ever had getting a job in my life. Nearly all the jobs I've had, I got real easy. Maybe I'm just lucky.

---

[1] I'll never forget Harold Glidden — the broadcasting consultant for WEGP. He was a great guy and a real pioneer. He built the first television station in northern Maine and was instrumental in putting the first radio station on the air in the Presque Isle area.

So there I was, hired. Instant Chief Engineer of WEGP. They had a 5,000-watt AM transmitter and a directional antenna array with two towers. Up to this time, I had never dealt with a transmitter of more than a thousand watts. But if you work on one transmitter, you've worked on them all. There's really not too much difference between one that gives out 1,000 watts and one that gives out 50,000 — except that the parts are bigger.

I asked Gerry what my hours were. He looked at me like I was crazy and said, "What do you mean what are your hours? You're the Chief Engineer — make your own hours." And I KNEW I was going to like this job!

Basically, the job of a chief engineer is to keep the signal on the air and keep it as good as you can. You keep everything within rules and regulations and the whole bit. You install new equipment (if you ever get it) and maintain what you've got.

I lived about 23 miles away from WEGP, but there was hardly any traffic on the roads of Aroostook County, Maine, so driving to work was a breeze. I got in the car and 30 minutes later, I'd be there.

The first thing I had to do when I started to work at WEGP was clean up the transmitter area. The transmitter room was a mess. It had everything from cobwebs and dirt on the floor to a wasps' nest in the tuning house. It needed everything.

When they hired me on, WEGP had just received a construction permit from the FCC to build a 100,000 watt FM station with a 500-foot tower on top of Mars Hill Mountain in Mars Hill, Maine. So part of my job was the installation of that facility.

I had never installed a 100,000-watt system before — but what difference does it make? It's just that all the pieces are bigger. WEGP flew me out to Nebraska to go to McMartain Industries and get some training on the new transmitting equipment. That kind of training is a joke. The only thing I really remember about that trip to the McMartain Industry plant in Omaha, Nebraska, was that on the day I was out there, the news came over the radio that John Wayne died.

So for a while, WEGP was my life. It was on 1390 AM with five kilowatts, omnidirectional in the daytime and directional at night. It was the #1 station in the market, and people as far as 30 or 40 miles away listened to it. If you wanted to listen to Top 40 rock, they were your only option.

Chapter Six
A Station Of My Own

As much as I enjoyed working at WEGP, I really wanted to get back into running my own radio station. So I contacted an engineering firm in Washington, DC, and asked them to find a frequency where I could run a decent power level. The engineers advised me try for a license on 710 AM.

In late 1979, I tendered an application with the FCC to build a 5,000-watt 710 KHz AM radio station on my farm. I wanted to make sure that my signal would be strong enough to cover Presque Isle, Monticello, and the surrounding communities. In northern Maine, you need a lot of power to reach your audience, because the population is so spread out.

Then one day when I was leafing through *Broadcasting* magazine, I came upon an ad from a brokerage company that sold radio stations. In one of their listings, it said they had FM stations for sale as low as $75,000. For the most part, that firm specialized in handling New England stations. And with the asking price as cheap as it was, I thought it just might be the little FM station at the record store in downtown Presque Isle.

I called up the brokerage company and asked, "Is WELF the station you have for sale at $75,000?" He confirmed that it was. I told him that I was interested in it, and asked him to send me some information. I read over the material and made them an offer of $70,000, which they accepted.

WELF was a small Class A station that was built in the mid-1970s by the fellow who owned the local music store. They operated on 101.7 FM and had a middle-of-the-road, this-and-that format. WELF ran 3,000 watts through a 3-bay antenna that was mounted on top of a 50-foot pole on the roof of the record store. But the antenna was so far down in the valley that the signal only went out about seven or eight miles. Sarah and I thought it would be interesting to take that little station and turn it into something. We didn't know exactly what when we bought it — but something.

When you buy a radio station, you have to apply to the FCC for consent to transfer ownership from one person to another. You can't just walk in there and buy it without the government's approval. So in the fall of 1980, I tendered another application for consent to transfer WELF from the current owner over to me. I didn't want to withdraw the application I'd submitted to the FCC to build an AM station up in Monticello — because if I had two stations, I could just simulcast one. Then I'd have a station on the AM and FM bands and it would be the best of both worlds.

The station was officially transferred to me on December 10, 1980. An engineering and site survey for a transmission site on top of Johnson Hill in Presque Isle was thrown in with the deal. It was a nice location and I knew it would give me a 15- to 30-mile range.

Once I bought WELF, the first thing I wanted to do was to move the station out of its location above the music store. I have this thing about paying rent. I really don't like it. So I looked around Presque Isle and found an old town house for about $23,000 that had been converted into apartments. It was a three-story apartment house — which was perfect, because I could keep the top floor apartment for my wife and myself, and use the first and second floors for the radio station. The house was located at 4 Second Street, right off Main Street in a commercially zoned area. We didn't have much money to put down on it, but the owners were willing to finance.

J.P. was in Maine at the time. I called him up and said, "We've got to go down to the record store and move the station." It was cold and windy outside that day — the temperature was around 8 or 9 degrees. The temperature inside wasn't very warm either. The guy that owned the building didn't believe in heat. The temperature in the radio station proper was about 50 degrees — which was another reason I wanted to move it out of his building. So in one day, we moved the entire radio station — the consoles, production studio, main studio, the transmitter — even the antenna.

After we moved all the equipment, I spent another couple of days installing the transmitter and installing a temporary single-bay antenna on top of the house. I wanted to eventually put up a 100-foot tower, but that was the best I could do for the time being. The main thing was just to get WELF back on the air and see what happened.

Shortly after I bought WELF, I made a request to the FCC to change the call letters. I hated WELF — it just sounded so diminutive. J.P. used to joke and call it "Elfin Radio." I wanted WOZ — "OZ Radio," but the FCC wouldn't issue three-letter call signs. So we had to pick a fourth letter. I asked J.P. about it, and he suggested WOZI. I applied for that, and it was granted in the early part of January, 1981.

But unfortunately, when I bought the station, all of WELF's advertisers canceled. We ran it for about two weeks that way, hoping that once they saw we had the station back on the air, the advertisers would come back.

At the time, I was still working at WEGP. Northern Maine is probably the only place in the country where you can build your own radio station and still work for the competition. Harold Glidden, the broadcasting consultant at WEGP, came up to me one day and said, "Allan, there's already enough rock and roll on the air. You ought to go country."

I said, "Country?"

He said, "Yes. Put on a country music format. You will get listeners. And they will be loyal to you for the entire life of the station. It's the only thing you can do." Some other people I knew in the area made the same suggestion.

I thought about it for a while and decided they had a point. There was too much of the same old rock and roll on the air. I talked to

Sarah about it, and she agreed. But I realized that I didn't know enough about country music to program it myself.

In the early 1980s, the programming services that are now so dominant were just beginning to take off. I contacted Tanner Music Services (which is now defunct) and asked if they had a country format. They said, "Yes — we just started a new country music format on tape. You can use it with an automation system (which plays the tapes automatically), or you can play it live or semi-automatic — whatever you want."

They offered to send us a few sample reels of country music programming to see how we liked it. Sarah and I sat down in the studio to listen to it, and we were totally impressed. At the time, country music was going through a renaissance. It still had the image of being twangy, cow pie-kicking music. But the newer stuff that was coming out was excellent. I said, "This is a winner. This is going to work!" And we immediately decided to subscribe.

Tanner Music Service cut us a very good deal. We ordered the basic library, which was about fifty or sixty reels of music. And every week, they sent us two 10-inch reels, which updated the music and kept the station fresh and lively.

The equipment that came with the original WELF was garbage, so I decided to temporarily shut down the station, put up a better antenna, and rebuild the facility. I got a new console, added two or three tape recorders, and basically redid everything. I put a 100-foot tower up next to the house and installed a better antenna on top of it. It was still a temporary system, but it was greatly improved. I also contracted to bring in the CBS Radio Network, because no one in the Presque Isle market was carrying it.

On February 2, 1981, WOZI went on the air as a brand new station. It was an immediate hit. Almost as soon as we signed on the air, people started calling us up and saying, "Thank you, thank you, thank you!" Or, "We're buying an FM converter so we can hear you in our car."

I wanted WOZI's announcers to sound friendly on the air and have a good time. I told them to develop their personalities, be who they wanted to be, and say what they wanted to say. The only thing I did insist on is that the country music format be maintained. And the music service tapes were kind of like a guarantee of that.

In commercial radio, consistency is important. People expect you to be what they want you to be. In other words, when you turn to a station on your radio, you go back to it because you know what it is, you know what they play, and you know what they're going to say to you.

The public loved it. But unfortunately, the business community was slow to respond. During our first year on the air, we sold spots for a dollar each just to keep the cash coming in so we could pay bills. I think our first ad was for a local auto junkyard. But even during those tough times, I never missed a payroll. However, I was on the extended payment plan with the folks at the telephone and electric companies for a while.

In the spring of 1981, we were granted a construction permit to move WOZI's transmitter site up to the top of Johnson Hill. I spent about a month putting up the building and tower. The signal went everywhere and covered all the major areas we needed to reach. Even the folks at Loring Air Force Base up in Limestone, Maine, could pick us up.

Our competition used to go crazy, because our 3,000-watt FM signal covered basically the same market areas as the 100,000-watt FM stations. And it really irked them. They even went as far as to accuse us of running too much power. But I just wrote to them and invited their engineers to come up to the transmitter and take their own readings.

That summer, the FCC granted me the permit to build WOZW, the AM station in Monticello. I started building the transmitter site in July or August. With every radio station I've built, there's never been enough money to do exactly what I want. I've always had to improvise and scrounge. This situation was no exception.

In the early 1980s, a new 5,000-watt transmitter cost around $30,000. But I was able to scrounge around and find 2 used RCA transmitters that Radio WTRY in Troy, New York, had up for sale for $700. They were close to 50 years old, but they worked!

I had a 20′ x 20′ x 10′ building erected on a concrete slab at the transmitter site for the unheard of price of $5,000. I asked the contractor to leave the front of the building open until I got the transmitters inside because they were huge things — about 30 feet long.

Back in the '30s and '40s when those transmitters were built, all the major manufacturers made oversized equipment. The transmitters were so large that you could walk inside of them and close the door behind you! Even a guy of my height — and I'm 6 feet, 5 inches tall — could go inside and stand up straight! Of course, they could be broken down into cabinets, but the pieces were still pretty large.

I had to drive all the way to Troy, New York, to disassemble the transmitters so that I could move them. In one day, I took them both apart. I separated the cabinets and took the panels, meters, and everything off. The local engineers in the area all came to watch and visit. They couldn't believe that anyone could get them apart so fast. But that's what determination does, I guess.

One sunny summer day, the truck came and dropped off the transmitters at my farm. Then the forklift operator and I spent half a day unloading 12,000 pounds of transmitting equipment. Everything was heavy — the transformers weighed about 1,000 pounds each and the cabinets weighed about 600 or 800 pounds apiece. It took me about a week to piece everything back together.

WOZW's 400-foot tower was the next thing on the agenda. It was delivered in 20-foot sections on a flatbed truck. My friend Harold Carmichael used his tractor and bucket to unload it. I thought he was going to ruin them all, but he didn't.

I painted the tower sections while they were still on the ground. Then I hired a company out of Portland, Maine, to erect it. It was quite a thing to watch it go up. They did a really good job. And about three days after the pieces arrived, a 400-foot tower was sitting on my farm in Monticello, Maine.

The last thing I had to do was put in a ground system. It consisted of approximately 120 400-foot-long radials, made of #10 gauge copper wires that were spun out at the base of the tower at 3-degree increments. I took an old farm plow and modified it to feed the copper wire down through a plow blade and bury it about six inches underground. I mounted a big reel of copper wire on the plow and hooked it up to my good old Farmall H tractor. It took me about a week to get it all done, but I really enjoyed that project. It was August, the weather was good, and it was nice being out in the sun.

All in all, the total cost of building WOZW-AM was only about $28,000.

WOZW signed on the air November 3, 1981. The station was a daytimer, so we were on the air from sunrise to sunset. Our first broadcast came from the studio at my house in Monticello. The next day, we began to simulcast it with WOZI. We were into duopolies before most people even knew what they were. The AM was a very good counterpart to our FM, mainly because people driving in their cars and trucks could listen to us all through Aroostook County. They could hear us from Madawaska at the northern part of the state all the way down to Bangor. That's the nice thing about low-frequency AM. Once you get below 900 KHz, the ground-wave signal really travels. If you have a good antenna system, 5,000 watts on 710 gets out as well as 100,000 watts on the upper end of the AM band.

By 1983, both stations were operating in full gear. But the advertising situation was still pretty rough. We had to do a lot of package deals and "dollar a holler" stuff. The economy in Presque Isle, Maine, was just like a lot of other small towns in the United States. Money was tight, and advertising dollars were spread thin. In New England, most markets had more radio stations than the area could economically support. So the only way a station could get advertising revenue was to take it away from somebody else.

After about two years of operation, we began to make some significant inroads. The local McDonalds in town conducted a survey, and we were one of the top two stations. That made us feel really good. At that time, the business community in general still felt that country music was a trashy format. And a lot of the businesses we contacted for advertising didn't think the people that listened to WOZI/WOZW had money to buy their products. But after the "favorite radio station" surveys came out, they started to realize the truth. People in the Presque Isle market liked country music — and they were listening to WOZI more than any other station.

Our listenership built every day, every minute we were on the air. And one by one, the advertisers started to come over. Soon, we were doing so well that one of the 100-KW FM rock and roll stations switched their format over to country. Everyone else at the station was pretty worried. But I wasn't all that concerned about it. I told them, "It just shows that we're doing something right — or else they wouldn't want to copy us."

True, the other station had 100,000 watts of power — but most of it was going out to the moose and deer. With our 3,000 watts, WOZI covered the market area we needed to cover — which was the Presque Isle/Caribou area. Fortunately, our advertisers and listeners stayed loyal to us. So after a few months of playing country music, our competitor went back to playing their rock and roll.

Before long, I decided that I wanted to do a talk show. So Uncle Fred's Just Plain Old Talk Show was born. Why Uncle Fred's Just Plain Old Talk Show? Because it sounds friendly. And radio needs to sound friendly if you want people to feel good when they listen to you.

We signed the talk show on every Monday through Friday at 10:06 AM — right after the CBS news. The talk show was open, and we talked about everything. I spent about five or ten minutes every morning reviewing the local paper for topics in case we needed them. But nine times out of ten, I had people calling in as soon as our theme song came on.

It's amazing how open-minded people in Aroostock County are. I know that some people who don't know the area think that Aroostook County is backward. But in reality, you find true independent Yankee thinkers here.

I soon discovered that most people in Aroostook County viewed government the same way I do — as a necessary evil we have to endure because people just can't be good all the time. But for the most part, the best role government can play is to just leave us alone and let us do our thing.

The drug issue is one thing we discussed on Uncle Fred's Just Plain Old Talk Show. I have always believed that the "War on Drugs" is nothing but a waste of money and resources. Some people just want to get high, and no matter what the government does, they're going to do it. So why not educate people, legalize it all, and set up rehabilitation centers with all the money the government is wasting on the "War on Drugs?"

I brought that subject up numerous times on the talk show. I painted the picture logically to people. I said, "Right now, the billions and billions of dollars people spend on drugs are all tax-free. The money doesn't go into the general welfare. It goes right into the pockets of some pretty nasty people throughout the world. So if governments legalized it and taxed it heavily, basically the same people that use it now would be buying it. The drugs would be clinically pure, and there would be less chance of someone accidentally getting an overdose. And more tax money could go towards educating people so at least they would realize what they're getting themselves into when they take drugs." Surprisingly, most of the callers agreed with my theory. And in time, I discovered that the listeners in Presque Isle and the surrounding areas were pretty open-minded about a lot of other controversial issues.

The early '80s was the beginning of controversial talk shows. Around that time, I started airing some stuff that really pushed the limits of what most stations were doing — within the limits of good taste, of course. This was when Howard Stern made his debut on WNBC 660 in New York. I remember the first time I heard him.

And it sounded to me like he'd been listening to my show and copied some of my material!

By early 1982, I was starting to get restless again. The Reagan administration's crazy genocidal maneuvers were causing needless suffering and death to people around the world. And I just couldn't allow myself to sit there and keep quiet about what was going on politically. I wanted very badly to provide a more powerful outlet than my local talk show for people to express their views over the public airwaves.

The 400-foot tower in Monticello, Maine, wasn't being used at night. And the thought of using it to radiate the signal of a "free" radio station was just irresistible. I toyed with the idea of taking a 300-watt transmitter, tuning it up to 1620, and testing it to see just how far it would go. So one day, I dug out an old WWII army surplus BC610 transmitter, which hadn't seen any real use since the WKOV days, and got it running. Then I fixed a switching arrangement so I could disconnect WOZW and put the 300 watt BC610 on the air.

I hooked up the transmitter, got it to match the antenna, and I called J.P. in Yonkers, New York, one winter's night. I said, "Turn on your AM radio and move the dial up to 1620. I'm going to turn on the transmitter and see if you can pick it up." I switched it on, and he could hear the signal with no problem at all. So I knew it was getting out at least 600 miles.

Within a week or so, KPRC Radio was born. It was basically an extension of what we did in the 1970s. People could call in and express their viewpoints with no delays or censoring. I know it was bordering on insanity to use a licensed AM radio station's tower after hours to do clandestine radio. But communications is important to me. In retrospect, it might well have been a very stupid thing to do. But we did it anyway.

On select nights, mostly during the winter, I threw the switch and turned KPRC on the air. Programming was fed from various studios in the New York area via the telephone circuit. That is when we began to experiment with equalizing phone lines. It's done every day now on the shortwave stations — but back then, we were just starting to experiment with it. We used only a standard voice-grade telephone circuit to get the programs from Yonkers, NY, up to Maine. The highs, if we were lucky, went up to about 3,000 cycles, and the lows went down as far as 150 cycles. But it still sounded pretty good. In fact, a lot of callers thought our audio sounded excellent.

Pirate Radio Central

KPRC   1616

We chose 1616 for our frequency because it seemed to be a good compromise. Nearly all AM radios could tune up to it, and we wouldn't interfere with any of the beacon services that were still on the air in some countries. It seemed to be the clearest, quietest channel. A BBC monitoring station even called in to let us know that the frequency we selected was absolutely one of the best for non-interference to any service worldwide. (In the two years KPRC was on the air, we never got even the slightest hint of an interference complaint.)

During the daytime, everything was normal. We ran our commercial country radio service. And then once or twice a month, depending on propagation, we'd turn KPRC on, usually around 10 or 11 o'clock at night. Most of the programs we aired on KPRC had to do with what was happening politically in the U.S.A. and around the world.

People called in on what they call "loop lines," where two people can be connected by calling in to a central tie at the phone company. It gives isolation to the studio. Every time we signed on the air, calls came in from all over the country. A lot of our listeners monitored the upper end of the dial every night, just hoping to hear our signal. The exchange of ideas was excellent. It worked out really well.

Sometimes, we ran KPRC until 2 or 3 AM. Actually, I was starting to get a little apprehensive about it. I didn't want the programs to run that long. I only wanted to broadcast for an hour or two. I thought that we were beginning to push our luck too far and we should try to cut things back before the FCC started getting interested. But it was hard to say no to the guys in New York — especially when they had so many important things to say about what was going on in the world.

In the winter of 1983, we decided to expand our coverage and put KPRC on 6240 shortwave. We ran about 100 watts into a dipole antenna. After that, we were operating on both 1616 and 6240 KHz.

We had some excellent broadcasts on KPRC. I didn't do any of the broadcasts myself. They were mainly done by J.P., Randall, and a few other folks down in New York who were into free radio. People called in with a lot of passion and emotion in their voices and talked about the crazy wars and conflicts that Reagan was getting us into. That was back in the early '80s when there was a lot of talk about the "Star Wars" nonsense, nuclear war, and blowing each other to bits. Reagan didn't seem to know what the hell he was doing, and there was a lot of insecurity about the world situation. KPRC was a good outlet for people to express their feelings about all this.

On the night of May 3, 1983, I begged the New York studio to go off after two hours, but they insisted on doing a four-hour show again. For some reason, I just didn't feel right about it. It was one of those premonitions I have from time to time. Then around 2 AM, a car came up the driveway. A couple of people got out, but it was so dark I couldn't recognize anyone.

I didn't even think twice about it. I shut everything off the air and jumped out the bathroom window. I just kept thinking of that Beatles song, "She Came In Through The Bathroom Window." But in this case, "He Jumped Out Through the Bathroom Window." I didn't know exactly what was going on, but I didn't want to take any chances. I got in my car and drove back home to Presque Isle. I figured that if it was the FCC, I'd find out about it soon enough.

The next day when I was up at WOZI, the doorbell rang. And sure enough, it was a fellow from the Federal Communications Commission — an agent that had probably been dispatched from the Belfast, Maine, Monitoring Station. He said that he'd picked up some signals on 1616 AM and thought it was coming from the transmitter site down in Monticello. I said, "Well, if anything like that came out of there, I can assure you it's not going to happen again." That was my answer and my statement. It was a very short conversation. I think he asked to see the station's license, and I showed it to him.

The agent left after about five or ten minutes. But then he came back and asked if he could go down and take a look inside my house. I refused. That was my private home. And without consulting an attorney, I really didn't want to take him through it. He said, "Okay, forget it," and then he left.

I was surprised because when the FCC does an investigation, they usually make a formal request to inspect the transmitter site — which under law you have to let them do if it's a licensed station. I went down and disconnected everything. I told the people in New York, "The FCC was here and it's over." And that was the end of

KPRC. It was fun while it lasted. Hopefully, we helped some people out and got them thinking about their environment and peace, love, and understanding — which has been the theme of every station I've put on the air.

It was good radio. In retrospect, doing it the way we did was not the wisest thing. But we didn't harm anyone or interfere with any other stations. People were helped. And a public service was done.

So it was back to the grind at WOZI/WOZW — playing that country music and cutting those commercials. In the early part of 1983, we actually did a lot with the stations. For one thing, we expanded to cable television. It's usually difficult to get a cable station to give you a channel or some audio space for your radio programming. But we were lucky. The fellow that managed the local cable system in Caribou and Presque Isle just happened to like country music. He also happened to be a really nice guy. So when I approached him, he said, "Sure — let me talk to corporate headquarters and see what happens."

The owners said he could do whatever he wanted. So we worked out a trade arrangement. We got Channel 20 on the local cable system in Presque Isle. And in return, we gave them unlimited advertising for whatever they wanted to promote for their cable service — the premium channels, movie channels, etc.

It worked out marvelously. To me, it was a perfect marriage. Our station gave the cable company a lot of publicity. And of course, having a channel on cable extended our coverage and provided another service to our listeners.

In the early days of cable, a lot of systems offered a weather channel — which was nothing more than a TV camera that scanned the analog instrument dials of a local weather station. It showed the rainfall, temperature, humidity, barometric pressure, wind speed, and wind direction. I bought the cable service's old weather system, hooked it up at WOZI, and put the instruments on the roof of the station. So when we weren't broadcasting video from the studio, we just put this weather station on the air so people would see a picture on the screen while they listened to us play country music. Some people told us they left their TV sets on all night long so they could watch the dials of the weather station — especially when it was windy and bad out.

I bought a portable color television camera and stuck it up in a corner of the studio. Then the DJ could switch himself on the air at his discretion. That worked out marvelously also. My two cats wandered in and out of the studio all the time, and people used to get a kick out of seeing them on television.

Our most popular announcer was a fellow by the name of Gary Stone. One of my cats always slept on the console table next to him when he did his afternoon show. People called in and requested that we switch the live camera on so they could see Gary and the cats.

Eventually, we started making our own videos to show on cable. Once we made a video about the welfare department of Presque Isle, and it got quite a reaction. Another time, we produced a video about space aliens coming to Presque Isle. It was all a lot of fun.

Back before TNN was on the air, we aired country videos. In the early '80s, there weren't all that many country videos around. But we contacted all the record companies and had them send us whatever they had available. Within a month or so, we had a library of a few dozen. When we played them, the audio went over the air and the video aired on Channel 20. That turned out to be very popular, too.

We also began to televise the Midnight Express show, which we aired every Saturday from midnight until whenever we got tired. That worked out pretty well. But we did encounter a few problems. One night, some weird and crazy people decided to come into the studio and expose themselves. In fact one guy did manage to moon the audience before we could shut the cameras off. We almost lost the channel over that. But I took care of the situation, and the cable company decided to keep us on.

In 1984, we did a lot of shows on George Orwell. I said, "Here it is, folks. It's 1984 and it's getting close." And we talked about all the things the government was doing to take people's rights away.

But I was starting to get frustrated again. Even though I was now living in Maine, I wanted more than anything to provide a radio service to my hometown, the city of Yonkers, New York. 200,000 people live there and to this day, it has no local radio station.

So one day, I was reading the FCC rules, which I do from time to time. Why not? I want to know my limitations. I want to know if I am in violation. And then I came to Part 74, known as the Auxiliary Broadcast Services.

Auxiliary Broadcast Services are used for a multitude of things — relaying programming back to the studio, transmitting program material and instructions to remote units, and as an emergency transmitter that interconnects the studio to the transmitter when the main studio-to-transmitter link transmitters die. They cover a wide range of applications.

Then I happened to notice that one of the frequencies allocated for the Auxiliary Broadcast Services was 1622 KHz. And I started getting excited. Then I carefully read the rules and the limitations. In fact, I even had an attorney check it.

Relating to the state of Maine, there was nothing in the rules that said an auxiliary broadcast station could not originate programming. At least, that was my interpretation. And I thought, "Hey! I see a perfectly legal way to put a radio station on the air in Yonkers, New York!" It was like a light went on in my head. It was one of those great moments in life when you see a way to do something that is proper and right.

I called up my friend, J.P., and I said, "I think I found a way to legally build a station in Yonkers, New York, on 1622." He was as excited about it as I was. So I immediately decided to apply for a license and a construction permit to build a 100-watt auxiliary broadcast station on 1622 KHz, with the transmitter to be located at J.P.'s mother's house, overlooking the Hudson River.

I got the application forms from the FCC, filled them out, and sent them in. To be honest with you, I didn't know if they would grant it or not. I thought that at least they'd want to know why a station in Maine would want to build an auxiliary station in Yonkers. But it all fit the rules, so I was optimistic about it.

About two months later, I received a letter from the FCC. I opened it up, and there was a license for a station: KPF-941, to operate on 1622 KHz and run 100 watts of power at 657 Warburton Ave. in Yonkers, NY! I was ecstatic. I went back to the station, called J.P. and told him that we had a license to go on the air.

The first thing I had to do was come up with a transmitter. I bought a 250-watt antique transmitter from a station in the Midwest for $100 and converted it down to 100 watts. What a great deal! It cost more to ship the blasted thing than it did to buy it.

I went to Yonkers in October of 1984 to build KPF-941. The transmitter took a couple of days to recondition, but it worked really well and tuned up excellent. It was a grid-modulated Western Electric transmitter, and it had some of the best fidelity I have ever heard from an AM transmitter. Again, I wanted everything to sound good and be proper.

KPF-941's antenna, a guyed 40-foot vertical, was made out of 10-foot TV antenna mast sections I bought at Radio Shack. It had a ground field of a dozen radials or so.

After I finished the transmitter site, I contacted the phone company and ordered a DC circuit to run from the studio at J.P.'s apartment at 70 Saratoga Avenue to the transmitter site at 657 Warburton Avenue. We needed two lines — one for control and one for the audio. They were essentially flat to about 8 Kc, so we knew that they would give us excellent audio quality. It only cost us about $25 to $30 a month for both lines, which was very reasonable.

We went on the air with KPF-941 on November 7, 1984. The signal did very, very well with our 40-foot antenna — which is electrically small for 1622 KHz, but adequate. It didn't sound like a booming 10 KW signal, but it was certainly listenable. KPF-941's signal covered the city of Yonkers and went up and down the Hudson River Valley.

We ran KPF-941 as a community radio station for Yonkers. The station aired local programming, and it was non-profit. The first two weeks we were on the air, we operated from 9 PM until around 2 AM in the morning. J.P. usually did one night and Randall did the next. We linked up with Randall in Queens, New York, via an equalized phone line. We played a lot of music and took phone calls from listeners.

At night, we had excellent skywave coverage. People were calling and writing us from all over the country. A lot of people called up just to find out what we were. We announced ourselves as KPF-941, because those were the call letters the FCC assigned to the license. But most of the people that heard us didn't know anything about the Auxiliary Broadcast Service.

We attracted a lot of attention locally. Lots of people were calling us up and the newspapers, magazines — even *Radio World* magazine — wanted to know what we were all about. We explained that we were an auxiliary station of WOZI doing some programming of local origination and transmitting them to members of the station — and anybody else that wants to listen.

After KPF-941 had been in operation for a few weeks, the FCC came up for a visit. At first, they didn't even know what it was. They saw the license, inspected the transmitter, and left.

But then I got a telegram from the FCC stating that we were to cease operations of KPF-941 immediately, which we did. I called them up and said, "Hey, what's the problem?"

And they replied, "You're operating KPF-941 illegally."

I had the rules right in front of me on my desk and I said, "Show me in your own rules where it says we cannot do what we're doing."

And the fellow from the FCC at the other end of the line said, "I don't care what the rules say, we're telling you that you can't do what you're doing."

It didn't make any sense. I was just flabbergasted. I said, "If you folks don't follow your own rules, and can't acknowledge your own rules and regulations, then how am I supposed to know what to do?"

I contacted my attorney and I said, "Look — I know we're not violating the law. What can we do?" He explained that we needed to run it more as an auxiliary station and not as a local broadcast station. So we went back on the air in a couple of weeks, operating it basically as an auxiliary station. In other words, all the programming that went over KPF-941 originated at WOZI, and was designed to be listened to by employees of the station that happened to be residing in Yonkers, NY.

I know it was kind of a round-about way, but it kept us legal. So we went on the air with KPF-941 in the later part of the winter of 1984/1985. We operated for about two weeks. Then one day I received a telegram from the FCC saying that they planned to revoke the licenses for my AM station, my FM station, and KPF-941 as well.

I remember the day it came. I looked at the papers and said, "I can't believe this! I'm trying to follow the letter of the law, I'm doing everything within the rules and regulations, and they're going for the jugular."

The government was saying that because of the so-called misuse of KPF-941 and the allegations of what had happened with KPRC back in 1982 and 1983, my licenses were to be revoked because I did not possess the character to be a licensee of the Federal Communications Commission. I had to hire a special FCC attorney in Washington, DC, to represent me, because this was a very serious thing.

This was one of the crucial turning points in my life, when I realized that it doesn't matter what the rules say. If the government doesn't want you to do something, you can't do it. If you try to do it anyway, they'll get nasty. They won't slap you on the wrist — they'll try to destroy your livelihood.

Needless to say, everyone was very upset — especially my wife, Sarah. She got extremely upset about it. And she was not only upset with the FCC — she was also angry with me for taking the risk of using KPF-941 to broadcast into my home town of Yonkers. It put a terrible strain on our relationship, and our marriage was never quite the same afterwards.

My lawyer summed the situation up this way. He said, "Allan, I know what you were trying to do with KPF-941. You're a little bit ahead of your time. You're kind of like the guy in the foxhole. The shells are blasting all around you (the shells being the government). And everybody does the normal thing in a foxhole, they stay low and just put up with the bombardment. But you're the person who likes to stick his head up out of the foxhole and take a peek every once in a while and just see what's going on. And you like to experiment. So every once in a while, you're going to take a hit."

Then he said, "Listen, they want to take your licenses and revoke them all. You have two options. You can fight it, and probably lose — or you can take the minority distress sale bailout."

I asked him, "What is that?"

He said, "Well, that's a little out that the government gives you in a situation like this. You can opt to sell your station to a recognized government minority for 75% of its assessed value."

And I said , "Do I really have a choice?"

He said, "No, you really don't have a choice, because if you go the other way, you'll probably lose all your licenses, as well as any money you put up to fight it."

The stations weren't doing too badly, but my financial resources were limited. That is another thing I learned real quick — when you take the government to task, plan on spending at least $50,000. And I certainly didn't have that kind of money.

So I said, "All right, I'll opt for the minority bailout." The papers were prepared and the government agreed. I had a friend who graduated with me at Ricker College who just happened to be black — Michael Carlos. I approached him one day and said, "Look, Michael, how would you like to buy my radio stations?" Michael was really into radio, and it was like he'd won the lottery. I explained what happened with the FCC, and he said if we could develop financing, he would do it. I asked my sales manager, Jeff

Franklin, to buy the radio station along with him because he knew the market and could help Michael keep the station on an even keel.

An agreement was drawn up rather quickly, but it took about a year for all the paperwork, financing, and everything to go through. In July 1986, WOZI was officially transferred to Carlos Franklin Communications.

In hindsight, it's easy to say that I shouldn't have done this or I shouldn't have done that. But what was done was done. I tried to do good with KPRC and KPF-941 — as I have with all my stations.

And life goes on.

Chapter Seven
Preparing to Broadcast
from the Sea

As early as 1984, I was thinking about putting a radio ship off the coast of New York City. At the time, Radio Caroline was on the air broadcasting full-force into Europe. And it was making money.

So I started researching the operations of all the existing offshore stations. I even contacted the United Nations and talked to some treaty experts about it. And from all I could gather, there was nothing illegal about operating a radio station off the coast of the United States on a ship with foreign registry. In fact, there was no direct wording about high-seas broadcasting in any of the international treaties back then.[1]

For a while, I didn't tell anyone that I was seriously thinking about building an offshore radio station. When I finally mentioned it to Sarah, she thought it was an exciting idea and that I should look into it. Soon after that, I told J.P., Randal, Michael Schaitman, and a few other people. They also thought it was a pretty exciting idea.

Once the decision had been made to broadcast from the sea, the first thing I had to do was come up with the funding to buy a ship

---

[1] To this day, I don't believe that there are any legal encumbrances to broadcasting from international waters into any nation, excluding of course, the United States and England. Those governments just won't have it. But on the high seas in international territory, I don't see how it could violate any FCC laws, British laws, or any laws that extend to international territory.

and outfit it with AM, shortwave, and FM transmitters. So I put together a business proposal and sent it out to numerous venture capital organizations and investment brokers. It explained how I intended to build the station and operate it as a free-based, free-thinking, open-microphoned alternative rock and roll radio station that would make money by selling ad time. I modeled my venture after what I saw the other offshore stations were doing — specifically Radio Caroline and The Voice of Peace, which was operating off the coast of Israel. The proposal included diagrams, charts, graphs, and a financial presentation.

It generated a lot of interest, but unfortunately, no money. I soon realized that the only way an offshore radio station was going to get built was through private funding. In fact, I almost liked that idea better, because I would be going in with a handful of people who seriously wanted to see it happen — and could draw from their expertise to make the idea work.

This was back in 1984 before the FCC decided to revoke all my licenses and destroy my business career. So when the FCC lowered the boom a year later, the idea of an offshore venture started to look really, really good to me. It seemed to be the only legal way to get on the air in New York, provide an alternative service, and make some real money at the same time.

In the early months of 1985, I decided to contact the Radio Caroline people. They had an office on Long Island, and I was somehow able to obtain their phone number. I called them up one day and said, "Hi, my name is Al Weiner, and I'm interested in working for you folks. I'm an engineer, and I can fix transmitters, build transmitters, and deal with all kinds of studio equipment. I know how to weld and I can do all the things that you have to do on board a ship. I'm familiar with diesel engines, gas engines, generator sets, propulsion units, pumps, piping, and all the other stuff that makes ships go."

Vince, the fellow who manned the office, invited me to come out and pay him a visit. So I went over to see him, and we spent the afternoon together. I told him that I had always been interested in offshore radio, and Radio Caroline was one of my ideal stations. He said I might be able to go over to the ship and work for a while, but he'd have to talk to some people before he could promise me anything.

Later on that year, I got to talk on the phone with Ronan O'Rahilly, the godfather/boss of Radio Caroline. He was very suspicious of me at first. He's the sort of chap who feels if he doesn't know you, you're working for the British government, the CIA, or maybe both. But finally, he decided that I was OK, and I got permission to come over and work on the ship.

I went to England on November 5, 1985, and stayed with some folks in a suburb of London called Finchley. I was there for about a week and a half until I could catch a tender going out to the radio ship — which at that time was about 12 or 13 miles out to sea near the Thames Estuary.

Under British law you are not permitted to service or take care of offshore radio stations. So to reach the ship, we had to go in a round-about way, leaving at night and all this clandestine stuff. The sea was really rough that night. Our little 60- or 70-foot fishing trawler was constantly pitching and rolling. We all stayed down below, and everyone was having trouble keeping their dinner down — if you know what I mean.

We finally arrived at the radio ship early the next morning. The *Ross Revenge*, the Radio Caroline ship, is a 210-foot cod trawler. It's quite a massive and heavily built vessel. Just as dawn was beginning to break, we came upon her. It was quite a sight.

The sea was still pretty rough. There was lots of wind, and grey skies. It took us about a half an hour just to safely get up on the lee side of the *Ross*. Getting on board meant timing the waves so that you could step from one rail to another without being smashed between the two vessels.

I spent about three weeks out there on the *Ross Revenge* doing all kinds of stuff — working on transmitting equipment, installing new audio-processing gear, and things like that. I really had a great

time on that ship. It was just like being at a land-based radio station
— except it was out on the ocean.

In 1985, the *Ross Revenge* housed two AM radio stations. Radio
Caroline put out a 5,000-watt signal on 558 Kc. They got into
London real well — the signal just barreled across the salt water.
And Radio Monique, a Dutch language service, had a 50,000-watt
signal on 963 Kc. They paid Ronan a lot of money every month to
broadcast their programs into Holland.

Life on the ship was very interesting. I got up at about 4 PM in
the afternoon and started my day. I checked around in the transmitter hold to make sure all the machinery was running, then I went
down to eat dinner. There was a cook on board the ship, and she
was really a nice gal. She cooked up one big meal a day. Everybody
met in the galley to eat at 6 PM — except, of course, the guy that
was on the air announcing. After supper, everyone sat around and
talked. For entertainment, we had television, lots of videotapes, and
plenty of books and magazines. Plus, of course, we had radio.

There were a lot of colorful figures on that ship. A fellow by the
name of Mike Barrington was the ship's engineer. He worked at
night and took care of the motors and diesel engines. He was a
character out of a Dickens novel. He had to be.

Besides me, the only American on the *Ross Revenge* at the time
was John Ford. He was a paid announcer who did programs for
Radio Caroline. I told him I was interested in putting a ship off the
coast of New York City and doing an operation similar to Radio
Caroline's. We hung around together a lot and talked about radio.

I had a blast — I really did. There was always something to do,
and there was always something to fix. During my time on the ship,
I reconditioned Radio Caroline's old shortwave transmitter. It
created quite a bit of interest when we did some international shortwave tests on 6.215 MHz. When nothing else needed to be done, I
put together some book shelves and tables. On the *Ross Revenge,*
there was a tremendous lack of simple things such as tables, places
to put magazines, and other amenities. So I gathered up bits and
pieces of leftover lumber and started banging things together. And I

wrote a letter to Sarah every single day that I was away because I missed her tremendously.

The weather was its typical gloomy self during my stay. The ride on the sea got a little rough every once and a while, but it wasn't all that bad. The ship would always pitch and roll. Then after about three weeks, a tender came out and took me back to shore.

But before I left England, I wanted to see Ronan. He agreed to meet me in a pub in downtown London, and we spent four or five hours talking about everything from radio to the death of John Lennon.

Ronan was very friendly with John Lennon and Yoko Ono. Ronan told me that a few days before John was killed, he called him on the phone and warned him that they really should be more careul. He believed that the CIA had a plot to kill them, and thought they should hire a bodyguard.

Ronan was adamant about it. He said, "You're an American and you can appreciate this. The CIA and the underground government of the United States is very, very evil. And they can do a lot of horrible things." Ronan believes completely that John Lennon's murder in 1980 was set up by the CIA — because at the snap of his fingers, John was able to rally half a million people against the government, for free speech, against evil, or whatever. The government saw him as a tremendous threat because he had that kind of influence and wouldn't play ball with them.

Ronan also talked to me about Loving Awareness — Radio Caroline's philosophy. It's hard to describe it. I don't even know if I can put it into words... it's more of a feeling. Loving Awareness is all the good thoughts and vibrations that you are particularly tuned into at the moment, a feeling that you can impart into the world around you.

My impression of Ronan is that he is genuine. His philosophy with Radio Caroline has always been to impart some greater level of awareness to the people who listen, as well as to the people involved with the actual broadcasts and running of the station. And for this, I take my hat off to him. He is an enlightened soul and a good man. He has definitely provided the world with a service. Radio Caroline was instrumental in opening up broadcasting to everybody — especially the pop music culture of the '60s and early '70s.

Ronan wanted me to stay on for a while, but I told him that I had a radio station to sell in the States and I was thinking about putting my own radio ship on the air. Ronan suggested that I call my station Radio Caroline West. There was a part of me that wanted to do that, but Randall, J.P. and the other people I was dealing with voted to keep our own individuality and call it Radio New York International — RNI.

When I got back to Maine, it was good to see Sarah again. For a couple of weeks, we got along pretty well. But then things started to fall apart again. And as soon as Sarah came home from work in the evening, she disappeared with her newfound friend, Rose. In the fifteen years I'd been with Sarah, she had never really had a close girlfriend. So even though I missed her company, I was actually pretty happy for her.

Sometimes I feel like destiny is pointing me in certain directions — and at this time, offshore radio seemed to be the thing to do. But I was still in the process of trying to sell WOZI/WOZW and bail out of that idiotic crap with the FCC. My sister Barbara took care of a lot of the management responsibilities. She did a capable job, but I still had to be there to run the station.

When you decide that you're going to build an offshore radio station, the first thing you have to do is find a suitable ship. At the time, a 36-foot sailboat was the largest boat I ever owned. Not knowing much about ocean-going ships, I had no idea where to look. I started doing a lot of research on all the ships that were used for offshore broadcasting.

I eventually decided that something around 150 feet, perhaps an old freighter or an old fishing trawler, would be adequate. I didn't really know what price range I was looking at — but I was hoping to find something under $100,000. Something I could perhaps finance or go into a joint partnership with. But I really didn't know where to start. I called some of the ship brokers that had listings in the back of my sailing magazines and asked where a person could get hold of an old fishing trawler or maybe a lightship. They told me I should check with the multiple-listing service. They charged $25 for a subscription, and sent me a whole bunch of listings on all different kinds of boats and ships.

Then someone told me that the best place to find commercial vessels for a good price is *Boats and Harbors* magazine — affectionately called "The Yellow Sheet," because it's printed on yellow paper. I got their address from a friend and I called them up. They told me to send in $12 for a year's subscription, which I did. *Boats and Harbors* had ships of all kinds and shapes — everything from garbage scows to luxury liners.

I started looking through it in earnest. In the classified ad section, I found an ad that read "150-foot fishing trawler for sale." I believe the asking price was about $100,000 or best offer. It gave a Boston, Massachusetts, phone number. So one day in January, 1986, I called them.

A guy named Frank answered the phone. When I told him I was interested in buying a ship, he said, "Great. I have two of them."

The first ship he offered me was a huge freighter — about 210 feet long. He wanted a lot of money for it, so I inquired about the 150-foot vessel. After we talked a while, Frank said, "Why don't you come down and look at it and make me an offer?"

So I called up J.P. and Michael Schaitman, and said, "Guess what — we're going to Boston to look at a ship!"

I stopped in Exeter, New Hampshire, to pick up Michael Schaitman, then we drove on down to Boston. I gave J.P. directions, and he came up from Yonkers to meet us. The ship was located at the Bang Boat Yard in East Boston, Massachusetts. When we got there, the wind was howling. It was 8 degrees Fahrenheit and the harbor was partially frozen.

At around 1 PM we went over to Frank's office — which was a shack off of Meridian Street. It was a one-room office, and everything was disheveled. There were old steering wheels, hydraulic pumps, generators, parts, nuts and bolts, and more, scattered all over his office. There were pinball machines, radar units, old soft-drink bottles, rancid coffee cups. The place looked like it hadn't been swept out since World War II. And there was Frank in all his jolliness, sitting behind his desk in command of his empire. It was just a typical waterfront atmosphere to make you feel nice and warm and cozy.

Frank is a well-built man. He's almost as tall as me, a powerful fellow with a real strong face, and a full head of hair — in other words, he's a very imposing individual. When he walks into a room, you know that Frank is there. He has a real deep voice, a commanding presence about him, and is a true captain in spirit. He's quite a colorful character.

We introduced ourselves, and Frank took us over to see the *Litchfield I*, which was over across the bridge in Peter Bang's yard. We had to walk out on floats — huge logs that are hand-hewn and strapped together to make floating pathways. Half of them were covered with ice and there were no guard rails. Then we had to crawl over a ship called *The Wave*, which was being reconditioned.

The first time I saw our future radio ship, she was covered with garbage. The people who were reconditioning *The Wave* had been using her to store scrap metal and wood while *The Wave* was being stripped down for conversion. Peter Bang promised us that if we decided to buy her, they'd clean everything off. So we got on the ship and looked around.

The *Litchfield I* was an interesting ship. In 1985, it was confis-cated by the Coast Guard off the coast of Boston. It had a 6-cylin-der engine running on 5. The main engine had blown one of its pis-tons, and it was blowing out tremendous amounts of black smoke. The Coast Guard stopped the *Litchfield I* because they thought it was on fire. Then they noticed that there were some unusual looking people on the ship. So they boarded it — and found out that the ship was loaded to the gills with marijuana.

The ship went up for public auction and Frank bought it for very little money. In fact, the money he spent for that boat could literally buy a week's worth of groceries. That's what can happen at public auctions. To this day I don't know the exact figure he paid for it, but I think it was under $100. I paid quite a bit more than that, but what did I know?

The ship was in rough shape. It was rusty, and paint was peeling everywhere. The engine room was a mess too — but I was basically looking for something that was cheap, seaworthy, had a good hull, and had space below deck where I could build a radio station and house generators and transmitters.

The ship had originally been designed as a refrigerated fishing trawler, and the piping and cooling coils were still there. The holds were large and they had high ceilings. Since the *Litchfield I* was a refrigerator ship, she had about a foot of insulation on all of her decks, sides, tops, and bottoms. As soon as I looked her over, I knew that the ship would be more than adequate. There was plenty of room for the studios, transmitters, workshops, places for people to sleep, and anything else that we decided to put on her.

But the ship needed a lot of work. When the DEA went through the *Litchfield I* and busted her as a drug boat, they tore up the floors and made a real mess of the place.

We discussed it and agreed on a price of $22,500 for the ship. I gave Frank a $5,000 deposit and agreed to take possession of the vessel on April 1, 1986. I was still going through the FCC proceedings with the minority distress sale of my stations, and my attorney advised me that it would be better to put the ship in J.P.'s name instead of mine. So that's what we did.

At last, I had a ship to build my station on. Happy with that, I went back to Aroostook County to continue operating WOZI, and started making plans to outfit the ship in the spring and summer of 1986.

Around this time, I started keeping a journal. Since these entries express the spirit of the time so well, I'll let them tell you the story...

April 23, 1986. I told Sarah that I bought a ship to convert to an offshore radio station, and she took it very well. In fact, she took it so well that I was completely stunned. Sarah seems to be genuinely excited about this. In fact, she wants to help. What a load off my mind that she is on our side.

April 24, 1986. Today Sarah hung a picture of the *Litchfield I* in her office. I am making ready to go and close the deal on Monday with J.P. and Michael.

April 26, 1986. At 2 PM, I arrived at the ship. J.P. and I were scheduled to close the deal with Frank at 4 PM. I puttered around looking at all the problems with the *Litchfield I*. At one point I sat in the radio room, looked at the mess around me and asked myself, "What am I doing buying this 400-ton piece of rust and corrosion? Maybe I shouldn't. But if I don't buy this ship, what other ships are there that can be afforded?" At 5:30 PM, I signed the sales agreement. I handed Frank a check for $17,500, and the *Litchfield I* was ours. Now the work begins.

April 28, 1986. Today Michael, J.P., and I went to the ship. We scraped, pried boards, swept, and picked up all kinds of trash. Michael poked around the engine room, turned a few valves, pressurized a few lines, and spun a few flywheels. J.P. measured the hold and helped where he could. I started to prepare the rooms for painting and general habitability. I also got the lighting systems on the ship energized. There is much rust, rot and corrosion everywhere. The holds are the best part of the ship. I cleaned and fixed up the radio cabin, Michael cleaned up the wheel house, and J.P. is trying to figure out how much lumber we'll need.

May 4, 1986. Today we rented a truck for $500 and picked up the 5,000-watt AM transmitter from WFST at Caribou, Maine. J.P. and I heaved and ho'd and finally got the one-ton transmitter on the truck. All went well despite the terrible cold and rainy weather.

June 11, 1986. Mike, J.P., and I all worked on the ship today. We cleaned out #2 hold. J.P. decided to put the studio in the transmitter hold. Mike and I started in the engine room.

(That was the last time that J.P. and Michael came out to work on the ship. Michael couldn't do it anymore because his wife didn't want him to be involved. J.P. was working at WFAS Radio and couldn't break away to spend time on the project. So after June 11, I was pretty well left to outfit the vessel by myself — save for the few times that John Ford, Randal and some of his friends came to help me.)

August 1, 1986. Summer is going fast. I had hoped to make more progress by now. Yesterday, I constructed a place for the studio. The inside walls went up and a console table was built. I also built a frame for the 5,000-watt AM transmitter and proceeded to start wiring it. I also connected a remote start switch for the water pump. And I finished the captain's cabin. Next week, I've got to get some help and come up with more money.

August 7, 1986. Today, I put up insulation and installed the outer walls of the studio. I've been trying to find someone to give me a hand one or two days a week. So far, it has been tough. John Ford is supposed to come to the ship next weekend.

All the driving back and forth is taking its toll. I am so very tired of driving all this way every week, but I have no other choice. I wish I could spend more time with my loving wife, Sarah, whom I miss so dearly.

(During the spring and summer of 1986, Sarah and I went through some difficult times in our marriage. I hoped that we could work things out. But before we could get our relationship back on track, Sarah fell in love with her girlfriend Rose and had an affair with her. I was hurt and disillusioned, to say the least.

(The many days we spent apart while I was working on the ship definitely put a strain on what we had left of a relationship. Sarah decided that the idea of building an offshore radio station was too wild, and she didn't want me to follow through with it. Many, many times I thought of quitting, selling the ship, and giving up. But somehow, I found the inspiration to go on.)

August 11, 1986. I just finished speaking with Randal. He is making up some test tapes to check out the audio and the transmitters. Next week, I'll have to go to New York to pick up the audio stuff from J.P.

August 15, 1986. A midsummer's night aboard the *Litchfield I.* I got here about 12:30 PM and installed the power lines and the main electrical panel. I had to burn off the feed-through pipe in the engine room, and almost burned myself up. Then I did a little bit of work in the studio. Now I am out of equipment to install. The FM transmitters are on the barge, the audio equipment is down in Yonkers, and the tower hasn't arrived yet.

(At that time, I had run out of money and was trying to get people to invest in the vessel. And you can bet your bottom dollar that I wasn't very successful. Most of the money came out of my own pocket.)

August 22, 1986. Monday I went to New York to pick up a load of studio equipment stored at J.P.'s. Tuesday I spent with Dad in Yonkers. Wednesday I left for Boston. I arrived at the ship and saw my bent and crooked antenna tower sitting on the crane barge. I was furious.

August 29, 1986. Today, I smashed my thumb assembling a piece of tower section. It has fought me every nut and bolt of the way all day long. Sometimes I wonder what the hell I am doing here. After some rest and dinner, it was back in the hold to work on the studio.

August 31, 1986. I finished assembling the lower sections of the tower. We all went out to Boston harbor to see the WBCN fireworks in Peter's rescue boat. I just had to take some time off from this work.

(Peter Bang is a great guy. He has got to be one of the most honorable men I've ever met. He does what he says, and never screws you. And in the waterfront ship business, that's rare.)

September 1, 1986. Today the 100-foot tower went up. It took about two weeks to get ready and 90 minutes to put it up with Peter's crane. Everything went smoothly. First, we swung the top three sections forward. Then Peter picked up the bottom two sections so John and I could bolt them together. Peter hooked onto the tower about 35 feet up — and up she went. The tower is now bolted into place. We strung the lower guys at 6:30 PM and finished up an hour later. What a job. What a tower. We are both beat. It was a long day, but now we have a real radio ship!

October 8, 1986. I spent the day on the ship. I finished installing the vertical FM antenna on the crow's nest. The Collins ring antenna for FM will go on the 100-foot tower. I'm getting ready to tune the AM transmitter. I'll have to start on the engine room soon.

November 12, 1986. I just realized tonight that I am a traveler. I travel from place to place in search of fortune. Today I traveled to Machias, Maine, to seek out a two-way radio telephone for the ship. Then I traveled to Augusta to spend the night with Pat and Linda before venturing on to Boston. A lot of miles have passed for me and my trusty steed of steel, the Dodge Aspen. We are both a bit more worn, but I know it is well worth the wear. My travels have separated Sarah and me, and that is the hurting part. We have both kept very busy and that helps. Life is short, and there are things to be done. It is cold out now, highs in the 30s. Tomorrow the ship will indeed be a freezer ship.

November 15, 1986. Today I entertained Randal and his friend John Calabro. They both loved the ship in all its rusty glory. I also painted a door and we cleaned out hold #2 in preparation for the ballast to be loaded in.

Dear Sarah, I always feel alone when I am not with you. Even when there are a lot of other people around. This project has taken its toll on me in lonesomeness. I arrived today in Yonkers at 6:15 PM, had some supper, took a shower, and watched *Dr. Who* on TV.

November 23, 1986. Life is a dream — so my grandma said as she looked back on almost 90 years of life in this reality before she passed on. Sometimes it seems to be just a dream, complete with real pain and joy.

Dear Sarah: I am back in Presque Isle laying next to you as you slumber and dream. I left the ship Friday afternoon after finishing painting the wheelhouse and piecing together the FM amplifier. Bad weather forced me to stop over at Pat's in Augusta. I am running out of work to do on the ship, mainly due to the fact that I am running out of money to buy any more parts and equipment — such as antennas, generators, and the like.

November 28, 1986. Today was a real fine day. Sarah cooked up the best supper I've had in a long time. On Wednesday, I found out that I can purchase a 60,000-watt generator for $1,000 from a boat school in Eastport. As soon as some money comes through, I will go down there and get it. It's a real break for the project.

(That winter, we were still getting everything together for the ship and deciding on a name for the vessel. Also, I managed to get WOZW turned over to a friend of the family, Dr. Rish. He is a Mexican/American, which is a recognized minority — so the government let me sell the station to him as part of the bailout. On March 5, 1987, Dr. Rish signed WREM (formally WOZW) back on the air as a talk radio station.)

April 16, 1987. The firetruck did a bang-up job of hauling the 60-kilowatt generator from Eastport to Boston. I must say that it was an exciting trip — hauling a 29-year-old surplus generator on a 43-year-old firetruck.

April 24, 1987. Today was rough. Fighting with pipes as I installed the lines for hot water. We had damp, wet weather. I've been banging my head everywhere on this tub. I started to tune up the FM antenna, only to discover that the interconnect line I ordered from Shively Labs was totally unusable. Today is best forgotten. It is very lonely here on this ship.

April 29, 1987. My heart is heavy. Yesterday I was in New York with J.P. talking to a fellow from a big advertising agency. He said he would only be interested in representing RNI if it had an X-rated format. That's right — it's the porn pirates or no deal. What has happened? Am I doing the right thing? Sometimes I feel that I am being driven by ideas that no one cares about anymore. Could it be that an offshore station in the U.S. wouldn't amount to anything unless it did a Playboy Channel of the radio waves? Do you have to insult, make fun of and describe sexual acts in detail to get an audience? I hope not. Love, peace, and understanding is what RNI is all about.

April 30, 1987. My wife has left me to be with her girlfriend, Rose. My father wants me to sell the boat. J.P. has no time. Randal is excited, but unpredictable. Michael is gone. What do I do? I want to keep going, but my heart and spirit are low. Give me strength, I must finish what I set out do. Goodnight, beautiful Mother Earth. Sleep well.

May 9, 1987. Randal, Danny, and Rhonda arrived in time to help me hoist the FM antenna up the 100-foot tower. Also, the 60-kilowatt generator and the hot water heater finally made it on board after sitting on Peter's barge for three weeks.

May 19, 1987. First, I energized the main electrical panel in the engine room. That went okay, except for an exploding switch. I energized all of the pumps and the air compressors, then I pumped up a bottle of air to 300 psi. Later I began pulling out floor plates to trace out where all the pipes and valves go. Lastly, I got the #1 generator to start. Despite the grease, oil, water, and rust, progress was made.

May 21, 1987. The engine room is killing me. My back is wrecked, my hands are sore, and my body is impregnated with oil. Today, I bought a 4½-ton anchor from Peter for $3,000. It's about a thousand dollars more than I wanted to pay, but it's a damn big and fine anchor. Plus, he had it right here in the yard, so there are no freight charges. With a 1½ inch bar-link chain, it will hold the ship well. I have so much more to do. I must continue and finish the vessel.

(This was a Japanese ship, and most of the ceilings were low and the quarters were cramped. And being that I'm 6 feet, 5 inches tall, I always had to work bent over — and it was really taking its toll on my lower back.)

May 22, 1987. Someone suggested that I register the ship in Honduras. Apparently, it's a lot easier and cheaper to do it there and we can even rename the vessel. I promised Sarah that I would name the ship after her, so I'll see what she says.

May 26, 1987. This boat yard is full of characters. Each one is a unique individual all their own. Lewis, a fellow from Canada, has been helping me lift heavy equipment around the ship. He is a great chap. He came to Boston to study at MIT, dropped out, and fell in love with boats. He's better off, I think — at least he wasn't turned into another mindless nothing building weapons for the government.

I spoke to Sarah. We might spend the weekend together on the coast of Maine. That news brightened my day. It gets so very lonely here aboard this iron tub.

May 27, 1987. To register this ship in Honduras will cost about $5,000. I hope I have enough money to finish. It's gonna be close!

May 28, 1987. This was a very busy day. First, I cut a hole in the generator hut to let more exhaust in. Second, I hooked up a fire-hose saltwater-pump system. Third, the bilge pump went in, and it kind of works. Fourth, I put more sealant on the cabin-house deck to stop the leak in the captain's cabin. Fifth, I installed a switch to switch between the Gates and Collins transmitters. Sixth, I put in a 15,000-watt dummy load antenna for the AM transmitter. Seventh, I am dead, bruised, beaten, sore and just generally exhausted.

Tomorrow I am heading home to rest my dilapidated body and do some on-the-phone work and prepare forms for the Honduran registry. I spoke to Randal tonight. He has definitely been a big help, taking care of all the programming and other related matters.

I'm tired, so tired now. I will cram myself into my bunk for another restless night.

June 2, 1987. I arrived back in Monticello. I spent the weekend healing from all my cuts and bruises. Monday I visited a few friends and picked up some fuel pumps and three fire extinguishers.

June 5, 1987. Back on the ship. I am somewhat rested. My back is better and my wounds have healed. This afternoon, I installed a fuel pump and a line into the transfer pump.

June 7, 1987. Today was spent battling it out with the old antique Gates AM transmitter. First, it had a power supply problem. Then I found out that all the modulator tubes are bad. I had to spend four hours tuning the bloody thing to the dummy load.

June 8, 1987. I debugged the RF out of the studio and tightened guy wires — generally a boring day.

June 9, 1987. Buying 400 feet of anchor chain will cost about $3,000. I am not happy.

June 10, 1987. Peter Bang is an interesting fellow. He can tell you the history of a Trico pipe wrench or why they taper the upper end of a machine bolt. This yard is full of masters of their trades.

June 11, 1987. I'm tired and my back is sore again. Today was spent debugging RF and tuning the Collins transmitter. (That was the 1-kilowatt AM transmitter we were going to use as a backup.) Fuel is going to cost around 55 to 60 cents a gallon, cheaper than I expected.

Tomorrow, I will be 34 years old. What have I done? No Sarah, no home, no income. What a confusing mess this is. I've got to get the ship out there and on the air.

June 16, 1987. It's just a damned shame that everything for the ship has to cost so much — chains, anchors, fuel and those crazy Honduran fees for registry. I had to buy a safety certificate that cost $740. John Ford got the stuff notarized and my father is running around depositing money in Honduran accounts. This is all crazy. It's insane — nuts! Prepare for the worst and hope for the best.

June 21, 1987. It's summertime now. The anchors on the ship went up and down as I got the windlass to work. Randal and a few of his friends are here helping out — cleaning the decks, pulling up planks and moving stuff. Plans are taking shape for the final push to New York. Frank is getting ready. There are still a thousand or so odds and ends to finish — stuff to order, things to buy. I'll most likely run out of money before it's all over. But as always, there are never enough funds to do just what you want.

June 23, 1987. I don't want to work. I just want to bang on the drums all day. It is a lousy cold day in Beantown. Fixing toilets, installing the pipes for the hot-water system — a real drag. I left the ship around 7 PM. I am now in Augusta, Maine, visiting at Pat and Linda's. Tomorrow I will go to Stonington and spend a day on my sailboat, the *Bob*.

I am spending all my nights alone now. Oh Sarah, what a time to switch polarity.

July 1, 1987. Much has happened since the last entry. Last Wednesday, I went to spend a day and a night on my sailboat, the *Bob,* only to lose the rudder in the heavy seas of a gale. If it wasn't for the CQR anchor, the boat would have surely been lost. Despite all the hair-raising adventures, I got the boat safely back to Burnt Cove by using a companionway hatch cover as a rudder. I arrived in Monticello on Thursday. On Friday, I spoke to a fellow in Honduras who told me they are now preparing the papers for the ship. J.P. should receive them in about a week. When it and the anchor chain arrive, the vessel could go out at any time.

Odds and ends are now being purchased. $1,400 worth of diodes, ring buoys, hi-fis, VCRs, chemical toilets, much more. All this is slowly eating away at whatever money is left in the account. I am afraid to even look at the balance.

July 5, 1987. Tomorrow, I leave for the ship. Maybe I can bring her down next week. The car is packed with ring buoys and a washing machine — plus a potpourri of assorted junk. If I had it to do all over again, would I? Probably so. I'd better get to sleep. It's breakfast with Sarah at 9 AM.

(At that time, I had reluctantly agreed to separate from Sarah because she wanted to stay in the relationship with her girlfriend, Rose. We were drawing up the separation agreement and dividing the property — so that's probably what our breakfast meeting was about.)

July 6, 1987. I arrived at the ship around 9 PM. The trash is still on board. The stuffing box is still stuffed, and Frank is down on Cape Cod with another anchor. Nothing was accomplished during my absence, as usual.

July 9, 1987. Down in Yonkers. Aside from the usual running around with J.P. and visiting with my father, not much is happening. But the papers did come in on the ship today. She is now duly registered as the *M/V Sarah* to the country of Honduras. Now we can take the vessel out any time.

July 13, 1987. Randal, John Calabro (Hank Hayes of the Brooklyn, New York pirate station WHOT), Ivan Rothstein, and Pete came up this morning to visit and work on the ship. Randal and his friends cleaned most of the main deck, and I continued to work on building a new head, which is coming along okay.

Peter will not go for my deal to buy his service boat to use as a tender, so that option is out. I just don't have enough cash. John Calabro's friend Pete has agreed to stay on this week and help me work on the ship. At last, a good helper willing to spend some time on this poor old rusty radio ship.

July 14, 1987. On Sunday, July 19th, the *M/V Sarah* may leave Boston bound for New York. There is still much to be done. Tying things down, locking the prop, and taking up the anchor chain using the windlass. Ivan, Randal, and John Calabro left tonight. Pete remained to give me a hand, and boy do I need help. Tonight we loaded ten bags of food on board, and there's much more to come.

I hooked up the washing machine today. Pete is making good use of it right now. Peter and I tackled the stuffing box. He put in a new ring of packing. (The stuffing box prevents water from coming into the ship around the propeller shaft that drives the boat.)

Randal is working on the programming. He brought a test VHS hi-fi tape on board, and it sounds great. The FM will sound good and clean. I am going to spend an afternoon trying to get the old tour boat alongside *Sarah* running. If so, I might drag it along as a tender.

July 15, 1987. Clear, sunny dry weather in Boston. I stuffed down the stuffing box, then spent the rest of the day working on the old Liberty Launch boat Peter wants to sell me. Tonight, Pete and I loaded another nine bags of food onto the ship. I called J.P. and notified him of our Sunday departure date. But I doubt if he will be coming down for the trip. Sad. John Ford will also not be on board because he had to go down to Florida and get a hair transplant. The ship's name, *Sarah,* is to be painted on the side of the vessel tomorrow. I must call Sarah and let her know.

July 16, 1987. Welding, moving chain and helping to move the boats around under the ship so we can paint the name "Sarah" on her. I am beat and sore. It has been double-time work trying to get this vessel ready for the trip. It's really nice to see how excited everyone is right before the big moment. I guess I'll buy that 40-foot tender boat of Peter's. Another wooden wallet.

July 18, 1987. Pete and I worked our buns off moving steel plates, mufflers, tires, and the like. There is a mad rush to get things done before our scheduled departure tomorrow. Yesterday, the *R.J. Munzer* came into the yard to pick up the anchor and some other stuff. She is alongside us now, ready to transfer fuel and tow the *Sarah* down. Randal, John Calabro, Perry, Pete, and a lot of their friends are now on board looking things over.

Alas, J.P. and Sarah are not here. All the people I love so dearly are not coming for the trip. Why? What have I done wrong?

(That was a sad thing. I really wanted J.P. to come down on the voyage, and he didn't. Sarah got stuck in some problems at her law office, so she wasn't able to make the trip either.)

July 19, 1987. No, we haven't left yet. Many, many, many things still need to be done before we're ready for the big haul down to New York. We should depart tomorrow morning after we hang about six tires off the side. During these last few days, we all have worked very hard. Tonight we transferred fuel into the #5 tank — about 2,500 gallons each. One of the tanks is rusted a bit at the top. There are now 5,100 gallons of fuel on board to power the generators — which I hope will last for a while. Randal, John, Pete, and Ivan seem to be having a good time. They have all been a big help. Peter is all paid off, and I bought the 40-foot Liberty Launch boat.

Oh Sarah dear, I am thinking of you. All I have is your name on this vessel to take me through. This ship wraps itself around me, and cradles me from the deep. I do miss you so.

July 20, 1987. Today we left. After hanging about 15 tires and Frank shouting at everyone, the *R.J. Munzer* towed the *M/V Sarah* out at about 1 PM. John and I took the tender boat and followed the two ships out.

The *M/V Sarah* looked great coming out of Boston Harbor this afternoon. The problems have started, though. First, the new head I built flooded with water. Then the day tank transfer pump burned out, and our 60-kilowatt generator isn't quite up to speed and is blowing out black smoke. But for better or for worse, here she comes!

July 21, 1987. At 7:00 AM, Frank got on the PA and yelled for me to go into the *R. J. Munzer's* wheelhouse to watch our ship enter the Cape Cod Canal. After many jokes about the tower not fitting under the bridges, we made it through. We are now at anchor off of the Cape Cod Channel.

The *Sarah* is moving along like a real ship now. Randal got seasick and is in bed. I fired up the AM and FM transmitters (of course, testing into a dummy load all the way). The AM transmitter put out a full 5 kilowatts, but then the driver burnt out. The FM worked fine. I have allotted a hundred gallons a day for fuel.

July 22, 1987. The *R. J. Munzer* dropped our anchor and chain around 8 PM this evening. What a day. We almost lost the tender trying to move the *Sarah* around to Frank's ship. Tomorrow, I start repairing, tuning, and getting the transmitters ready to broadcast.

We are now at anchor 4½ miles off Long Island, NY. One and a half years and $72,000 later, here we are! We have 30 days of fuel, food, and drinking water on board. Now we'll see if this idea works.

Chapter Eight
Radio New York International
Is On the Air

On July 23, 1987, Radio New York International was finally ready to broadcast from the good ship, *Sarah*. Our first offshore test transmissions went out on 103.1 FM, 1620 AM, 6.240 shortwave, and 190 KHz longwave.

Randal, John, and Pete went back to shore in the tender that morning. So that left Ivan Rothstein and me to run everything and take care of the ship. Ivan was pretty overwhelmed by the whole thing. I don't think he'd ever seen the inside of a real radio station before. When everybody else left the ship that day, Ivan was the only one who volunteered to stay on and crew with me. So he got to make radio history when Radio New York International went on the air to make our first test transmissions that night.

There's no way to describe what it feels like to be broadcasting, sending out signals of music and hope, while you're sitting in the belly of a vessel as she gently rocks you back and forth as the water cradles her in the great Atlantic Ocean. And after all those months of hard work, having Radio New York International on the air felt fantastic.

I can clearly remember that wonderful evening. It must have been 85 or 90 degrees outside, and all the fans were running. From the studio, you could always hear the constant thumping of the ship's generators and the hum of the transmitters. The place was definitely alive.

I let Ivan take the controls while I went out on deck. It was a clear night, and the tower's silhouette looked beautiful against the starry sky. The ship was gently rolling, and you could see the tower moving against the blanket of stars. It was one of those great unforgettable moments in my life.

Randal, J.P., and everybody connected with the project was listening on land. The 1620 AM and 103.1 FM transmissions came into New York City the best. We broadcast for about six hours that first evening. I did have a few problems with the AM transmitter's driver section. Still, it was a wonderful, incredibly exciting night.

But it wasn't long until the problems started. The next morning, I was doing some welding and all of a sudden, the power went off. I looked up and saw black smoke everywhere. The fuel in our 60-KW generator had caught fire. Frank saw the smoke and came by with the *Munzer* to help me put it out. The generator was mounted in a steel box, so there wasn't any ship damage. But after the fire, I had to go back to using the ship's main generator, an old Yannmar that was so old and decrepit that it barely ran.

Still, we managed to go on the air Friday night at 7 PM for another test broadcast. Then around midnight, the AM transmitter went out. At 12:45 in the morning, the water pump on the main generator sprang a leak. By 3 AM that morning, I was having to use the 300-watt portable auxiliary generator just to keep the lights on.

I knew there would be problems, but I never expected to lose so much in one day. That's the thing about ships. If one thing goes wrong, it can trip off a whole cascade of events that can cause many, many things to go wrong down the line. That was quite a day for me. We really had to boogie around.

The Coast Guard came by on Saturday morning, buzzed around for a while, and left. And surprisingly, no news people had come out to interview us or see what we were doing. It was just the sea, the ship, Ivan, and me. The temperature was about 90 degrees with 100 per cent humidity. Ivan and I spent most of the day lounging on deck, swatting flies, and reading old broadcasting and *Popular Communications* magazines, just waiting for the night to come so we could sign the station back on.

But then about 4:30 PM, things got real exciting in a hurry. The Coast Guard cutter, the *Cape Horn* (a 90-foot patrol boat), came out and started circling the *Sarah*. Her captain hailed me on the two-way and asked if he could board the vessel. At first, I declined. Then he came back and said that he just wanted to do a safety inspection. So I decided to be neighborly and let him come on board.

But little did I know that it was not only the Coast Guard — the FCC, the DEA, and the U.S. Customs Service were also with them. And once they got on the ship, they acted like rats on a feeding frenzy. They went over every part of that boat looking for drugs, hidden illegal immigrants, or whatever.

One fellow was extremely arrogant and nasty. He spent nearly 45 minutes continually hurling obscenities at me. After about five minutes of being cursed at and screamed at by this agent of the DEA or Customs, I realized what he was doing. He was trying to incite me to do something — get angry, take a swing at him, or start running. Anything he could use for an excuse to beat the living daylights out of me or maybe even shoot me. I guess that's their *modus operandi*.

I recognized that early. So I sat very calmly up in the wheel house of the ship and I didn't say much of anything. But this guy from the government just kept on threatening me and saying that they were going to impound the ship and destroy this and destroy that and arrest this and arrest that. I knew that we weren't doing anything wrong or violating any laws — if we were, they would have done those things already. But I felt very vulnerable and was concerned for Ivan's safety.

After about 45 minutes of constant harassment, this guy from the government started asking me questions. First, he wanted to know where the ship departed from. I was hesitant to answer him because by then, I was pretty annoyed about the whole matter. The Coast Guard had made it sound like they just wanted to make a friendly safety inspection — and being a good citizen of the sea, I allowed them on board. But instead, it turned into this gigantic government inquisition.

But after I thought about it for a few minutes, I decided that I'd better answer him. I turned to this officer that had been constantly cursing and screaming at me and said that the ship came from Boston and there were only a handful of people on board (whom I did not mention by name). He seemed content with that information because he stopped his attacks, then stood up and walked away.

An agent from the Federal Communications Commission was also on board making all kinds of threats and issuing another idiotic FCC warning. I repeatedly told him that we could not be in violation of the Communications Act of 1934 because the Communications Act does not have jurisdiction over vessels of foreign registry in international waters. This I knew and this he knew — even though he refused to acknowledge it.

When I asked him which rules and regulations he thought we were violating, he said, "That's not for me to determine, I'm not an attorney." Right then and there, I knew that this conversation was useless because he had no idea of what he was talking about.

So I left him to do whatever he wanted to do — which was look around the radio station and take notes. I went up to the galley to get something to eat. And I just kept thinking how big of a mistake I made in letting these people on board. They didn't do any actual harm, but still, I had subjected myself and Ivan to all this unnecessary, unlawful harassment.

The FCC agent tried to hand me a written warning stating that I was in violation of Section 301 of the Communications Act. I refused to accept his piece of paper, and he got very arrogant and angry about it. The last thing he said to me was, "Enjoy your Cheerios for dinner, because that's probably all you'll have to eat," — which was a weird comment to make. Then he took all his papers and left the ship.

After about an hour and a half of harassment, they were all gone. And I was still pretty upset about it. We were out there trying to offer an alternative radio service, not hurting anybody. And we get assailed by all of these government officials who seemed to be hell-bent on destroying everything for no good reason whatsoever.

Radio New York International remained off the air Saturday night while we decided what to do. But after thinking it over, Ivan and I were determined to continue with our broadcasts — because despite the government's threats and harassment, we believed we had every right to do so.

That day, I wrote in my journal, "At first, I thought very strongly of having Frank come back and pull up the anchor and bring the ship in. But after some soul-searching, I decided to stick it out. The jurisdiction of Section 301 of the Communications Act ends three miles out from shore. After all this work, I believe that this project can be a success."

On Sunday, July 26, Ivan was down below in the radio hold asleep and I was up in the pilot house looking out the window. We were still in the midst of a heatwave, and we were waiting for evening to come so we could make our next broadcast.

At around 1 PM, the sky got dark very quickly and the water turned from blue to green. It happened so fast that I barely had enough time to run around the ship and batten down the hatches. Then the lightning and thunder came — and it rained like I'd never seen it rain before.

The wind was forcing the rain against the glass with such intensity that I could barely see outside. We were getting wind gusts from 60 to 80 miles an hour. Pieces of lumber, chairs, and everything else on the deck of the ship was literally being picked up and blown overboard. For a while, I was really starting to wonder if the tower would hold up.

But after about 10 minutes or so, the storm passed. It left just as fast as it came. I went out on the deck and, boy, was it clean. Anything that was loose — boards, deck chairs, everything — was blown off. God knows where it all went. There was no damage to the ship, however. Ivan came up a couple of hours later and wanted to know what happened. I said, "Well, Ivan, you really missed it this time!"

We went back on the air that night and did another broadcast. My journal entry for that day read:

"Tonight, we went back on the air for another six hours. Who knows what the authorities will do next? We are breaking no laws, but they're treating us like criminals. The news media is finally starting to cover the station, and most of the New York papers are carrying the story. I sure hope we can get support and advertisers soon."

Monday, July 27, was media day on RNI. No less than nine reporting crews came out by boat to do stories on us. Ivan, Randal, and I were busy conducting tours and giving interviews. We had people from CNN, CBS, NBC, ABC, WPIX, WOR, WNYW, *A Current Affair*, the *New York Post*, and the *Daily News* — plus a reporter from the *Village Voice* who was staying on board to write an in-depth story.

We were really surprised at the press turnout. As soon as one crew finished videotaping and shooting everything, another crew came on board and did the same thing. Randal came out with one of the newsboats, and he stayed on board until Monday night helping us to deal with all the press people.

That was a good day. It was another hot day, but instead of swatting flies and reading magazines, we were busy with the news media. Most of the media people that came out were extremely friendly, and we had a lot of fun. We went on the air that night and had another great broadcast that lasted six hours or so. Again, we were just testing our equipment. So far, there was no sign of the FCC or military. Hopefully, there would be more radio to come.

On Tuesday morning, July 30, I was resting in my bunk and the ship was gently rocking back and forth. I was half awake and a cool breeze was coming through the porthole. But this lovely marine scene was soon to be shattered. At about 5:30, I heard a vessel coming up. And they started shouting at our ship, "All crew aboard the *M/V Sarah* prepare to be boarded!"

I woke up in half a daze, looked out the porthole and saw the *Cape Horn* full of Coast Guard men with rifles and guns. They had a couple of high-powered machine guns mounted on deck with men standing beside them. I just stood there and thought, "Well, I guess this is it. They're the government, they have all the guns — so what can I do?"

I got on the ship's radio and I called the captain of the *Cape Horn* to find out what was going on.

He said, "We are going to be boarding you."

I told him that we hadn't done anything wrong, and boarding our ship without just cause was against international law.

He acknowledged my objections, but he said that he had orders to board the ship and arrest everyone on board. He implored me that for safety's sake, I should get everyone out of their bunks and up on the deck.

At that time, the only people on board were myself, Ivan Rothstein and R.J. Smith, a reporter from the *Village Voice*. R.J. was already up, but Ivan was sleeping down in the radio hold and couldn't hear anything. So I asked R.J. to go down and wake Ivan up and have him come on deck.

With all these people toting guns around and pointing them at the boat, I was just a little bit nervous that somebody might get trigger happy and start shooting. Remember, our ship was in international territory, and that's a little different than being on U.S. soil.

The Coast Guard boarded the ship, and said that we were under arrest for breaking international law. The first question I asked was "Excuse me sir, but what international law are we being arrested under?" The Coast Guard man promptly stated that he did not know which law it was, but it was some international law.

Everybody was handcuffed. Then we were ordered to the forward part of the ship. The sun was rising. It was going to be another one of those 90-degree scorchers. And for six hours, we had to sit out in the baking sun while all those people from the government were running around figuring out what to do with the vessel.

The FCC had a party. They were running around taking videotapes, snapshots, and Polaroids. They cut all the cable harnesses and transmission lines with hacksaws, ripped everything out, and threw it in the middle of the deck. And of course, they damaged quite a bit of the equipment. They basically destroyed everything that it took me a year and a half to put together. The FCC knew that we were outside of their jurisdiction and were not violating the law. But they wanted to make damn well sure that we didn't go back on the air after they left.

I asked the Coast Guard, "Why are you letting these people do this? We have violated no laws, and you've come on board, seized our ship — as far as I'm concerned, illegally — and are letting the FCC tear up our station. Why are you letting this happen?"

One of the men admitted that he didn't understand why the FCC was destroying equipment and breaking stuff up. He didn't feel that was right, but he was under orders.

My main concern at the moment was everyone's safety. I told the Coast Guard that Mr. Smith was just a reporter and did not have anything to do with the radio operations. But they didn't believe me and they wouldn't allow him to show them his press identification.

While this was all going on, a flotilla of boats came out from Long Island. Of course, the press and television crews were out there — as well as curious onlookers who were trying to see what was happening. The Coast Guard had a patrol raft spinning around to keep curiosity-seekers from getting too close to the ship. A lot of people were yelling words of encouragement to us, and that made us feel a little better.

Ivan kept saying, "How can they do this? What did we do wrong?" I told him that the best thing to do was be quiet and not say anything. When we get back to land, I said, we can get an attorney and deal with this in court.

My concerns again were for everyone's safety. The important thing in a situation like this is that no one gets hurt and that no one does anything that they shouldn't do. After all, what can you do when government people come on your ship with guns pointed at you — and all you did was operate a radio station?

I requested numerous times that we be given something to drink and eat, but to no avail. Finally, I requested that we be taken out of the bloody damn sun, which they did. They brought us onto the *Cape Horn* and put us down below in one of the receiving areas. I asked the captain if we could have some food and water, and he had some drinks brought and sandwiches made. Then they moved our shackles to the front of our bodies so we could eat.

R. J. Smith was released at that time. After all of our pleadings, they finally realized that he was just a reporter and didn't have anything to do with the operation of the boat or radio station. The poor guy took it very well, but as the acting master of the vessel at the time, I felt responsible.

Ivan and I were transported on a 40-foot police boat to Brooklyn, where we were to be arraigned under Federal charges of violation of international law. We did not violate Section 301 of the Communications Act. They couldn't nail that on us. So they trumped up some charges having to do with violation of international law — which to this day I don't fully understand. They also charged us with some crazy nonsense about conspiring to impede the function of the U.S. government — whatever the hell that means.

A couple of FCC agents were on board the patrol boat. One of them was an agent by the name of Judah. He looked at me sitting there in handcuffs and started to talk to me. At one point in the conversation, he said this was not the way he wanted it to turn out. I took that to mean that even he was surprised at the government's overreaction to the situation. From what I could tell, if it had been up to him, Judah did not intend to go out there to arrest everybody and destroy everything as the government had done.

Judah said that he thought the radio ship was impressive. I looked at him and asked, "Well, why did you guys go out there and destroy it?"

He never really answered.

We were taken back to a Brooklyn pier and escorted off the ship in handcuffs. Quite a mob was waiting for us. Of course, the press was there. They asked me what happened, and I yelled out to them, "Free-form rock and roll radio had been snuffed out by the United States government!"

Lots of free-radio supporters were there to cheer us and wish us well when we got off the ship. That felt really good — especially after a day of sitting in the hot sun and wondering what was going to happen to us next.

Ivan and I were put into a police car and driven to the Brooklyn Federal Courthouse, where we were taken to a holding cell. We didn't have an attorney, and they would not permit us to use a telephone. So we were in contact with no one and didn't know what was going on.

When we were brought to the basement of the courthouse, I was pleased to see that Margaret Mayo, a partner in Dudley Gaffin's law firm, was there. Other than all the police officers, she was one of the first people I saw. Margaret told me she was willing to handle the case on a no-fee basis, and four other attorneys had come who were willing to do so as well. And I said, "Great!" because we were very vulnerable at the time, and we definitely needed legal representation.

We spent about an hour in the cell talking about what had happened. She said that we would be arraigned before a federal magistrate and he would decide whether or not a bail would be set. It was quickly determined that bail would probably not be required for Ivan, but it might be for me because I had had some problems with the FCC back in 1971.

The police officers at the Brooklyn Federal Courthouse were pretty nice to us. They allowed us to stay in a reception area outside of the cell block and did not lock us up. I got the impression that they just couldn't understand why anybody would be put in the detention cell — or thrown in jail — for operating a radio station.

Remember, most cops in that area are used to dealing with hard-core drug dealers and dangerous criminals. And to be handed a couple of guys that just got pulled off of a boat for the alleged crime of playing rock and roll music on the radio was probably not very high on their list of bad things people do.

Margaret told us what was going to happen and what the magistrate would probably do. While she was talking with us, I kept thinking, "How did I get from running a radio station of love, peace and understanding to this cellblock of the Brooklyn Federal Courthouse?" It was just tearing me up inside.

In about an hour, we were brought up to the courtroom — and it was really packed. There was standing room only. Reporters from all the local media had come, as well as a lot of free-radio supporters who had heard about what had happened.

We were brought before the magistrate. The entire case was apparently built on the testimony given by one FCC agent. He claimed that we were violating international law, obstructing the government and all kinds of crazy nonsense. All the citations on the case were done by him personally.

The magistrate read over the documents, then looked up at the agent and asked, "Sir, do you realize that you're not only saying this for the government, but you're saying this personally?"

He said, "Yes."

Then the magistrate turned to me and said, "Mr. Weiner, would you agree to remain off the air until the hearing date is set on this case in August?"

I said, "Yes."

The government wanted the court to require about $50,000 bail for me, and they didn't ask anything for Ivan. But the magistrate overruled them. He didn't see any reason whatsoever for me to post bail, and he released us both on our own recognizance.

That was the end of the arraignment. People were cheering in the courtroom and patting us on the back. When we walked out, John Calabro, Perry, and Randal were there to greet us. After all we had been through, I was very, very happy to see those fellows.

I had been trying to deal with the fact that we were probably going to be thrown in a jail cell for a couple of days until all this got sorted out. But that didn't happen. I was pleased, but I was surprised.

We all went out to get something to eat. We had a nice meal at a local diner, and I told the guys about everything that happened. Then we began to discuss strategy. Of course, the first thing on my mind was what would happen with the ship. I didn't know what the government was going to do with it, when I could get it back, and how much work I'd have to do to piece things back together. A few days later, I wrote in my journal:

July 30, 1987. Special offshore supplement — On Tuesday, July 28, we were boarded. Ivan, myself, and R.J. Smith were arrested, and the ship was seized. The FCC came on board and tore apart the radio station. It hurts to talk about it. I've tried very hard. I wanted to bring free radio to New York. Please, no failure this time. We need a solution. So many people have hoped and dreamed. To let them down is a pain. I guess it's part of the learning process. Love, peace and understanding.

In the following days, I spent a lot of time sitting in Dudley Gaffin's office talking strategy and dealing with the brigade of news reporters, well-wishers, and so on. The media's reaction to all this was much more than I expected.

The first night after the ship was busted, I stayed at John Calabro's apartment because I was just too tired to travel. It was quite an exciting night. The phone was ringing constantly, and all my answering machines were clogged with calls from people all over the country. Friends that I hadn't spoken to in years were calling up to find out what was going on and wishing me the best.

We got phone calls from all kinds of people. One of David Bowie's people even called to wish us good luck and say they thought that what we were doing was great. The support was there. The calls were positive and newspapers and magazines were constantly calling us for interviews. Randal was also busy doing interviews, making statements, and things like that.

The morning after the arrest, I got on the subway and went back to Yonkers to visit my father. (My car was still up in Boston because I left it there when we took the ship down.) Dad had just come back from vacation and what a thing to run into — all his friends calling him up and saying that his son had been arrested because he had a radio ship out on the high seas. But he was glad to see me and know that I was okay. Dad was always there and always concerned. He was just great about it. Then I went over to J.P.'s apartment and slept on the floor in my sleeping bag.

At the time, Sarah and I were more or less separated. She was still having the affair with her girlfriend, and that was a heavy thing for me to deal with. I still loved her very much, and I was sorry we had drifted apart. Everybody in Aroostook County knew about the ship, that I was arrested, and all this stuff that happened out on the high seas. The newspapers, radio, CNN, and all the major TV networks picked it up. The local television station WAGM even did a special news story on it when it happened. So I knew that Sarah had heard about the incident, and I wondered what was going on in her mind.

Then that night at about 2 AM, Sarah called me. She sounded slightly hysterical on the phone. Sarah was somewhere in Yonkers, and had driven all the way down from Maine to see me. I was really shocked. I tried to determine where she was so I could give her directions. About an hour later, she arrived at J. P.'s apartment on Saratoga Avenue. We slept together on the floor that night. After being alone for so long, it felt so good to hold another person again. We didn't "do anything" — we just slept there close beside each other.

But when we got up the next morning, Sarah seemed a little different, a little bit colder. She said, "Well, I just wanted to see you and make sure that everything is all right. But I think I'm going to leave now."

I said, "Listen, why don't we do this — I'll drive your car to Boston and pick up my car, and we can be together that long before you head back to Maine?" She agreed, and we had a very pleasant ride that day. She dropped me off at the Bang shipyard, and I picked up my car and said good-by. Then she drove on back to Maine to do her job at the law office and be with her girlfriend, Rose.

Sometimes, when I think about that day, I wonder why it happened. We had loved each other for such a long time — so I guess when Sarah heard I was in trouble, she wanted to be near me. Even though it didn't lead to anything, it was nice and I appreciated it.

After Sarah left, I stayed in the shipyard for a while. Everybody wanted to know what happened. I ran into Peter and told him that he'd probably have to pull the ship in because the government damaged it and one of the generators had burnt out. I picked up my car and headed back to New York.

Here are a few more of my journal entries:

August 4, 1987. Here I sit on my crippled ship, both generators are out, and the radio station wrecked. We did get nationwide media coverage. We're even going to be on MTV. The press is great — but sadly, Radio New York International is off the air. The ship is damaged, and so am I.

Was it all worth this so far? Most people say "yes." But my heart hurts — and maybe says "no." The great experiment in radio continues. Frank is supposed to tow the ship back into Boston and put it in Peter's yard. There it will sit until this legal stuff is over with. Maybe we will sue the government for damages, maybe we will get licensed for land-based operation, and maybe nothing will happen.

Anyway, here I sit on this broken radio ship with a heavy, heavy heart.

August 5, 1987. The *Munzer* is towing the *Sarah* back to Boston. What goes around comes around, I guess. Frank showed up yesterday afternoon and lifted the anchor chain last night. I am sorry that I had to move the ship from its offshore anchorage, but leaving her out on the sea is quite impractical. Both generators are out and the legal hassles could last a very long time.

Getting RNI back on the air is my #1 priority. Should I sell this rusty old vessel and get another? Maybe I could try to get hold of the *M/V Communicator* (the Laser Radio ship, which was currently somewhere in England). I'll give Vincent a call when I get ashore. The legal battle could last a while, especially if we sue the government for damages.

Randal, John, Pete, Ivan, Perry, and the others are now broadcasting on WNYG in Babylon, NY. They're doing a special "RNI Take Over" program.

(WNYG was a 1,000-watt radio station out on Long Island, New York, that broadcast on 1440 AM. They contacted us right out of the blue and said, "Look — why don't you come out here and do some programs from our station?")

Boy, do I have lumps on my head from banging it into just about everything on this ship. It's 11:10 AM — lunch is in an hour, and beer break comes at 2 PM. Peace, love, and understanding — that's what RNI is all about.

I must get RNI back on the air. It feels so strange to see the *Sarah* being towed back to Boston. All that work and hope for only four days of broadcasting. Was it worth it? We shall see.

Dear Creator, center of it all — I am tired and very dirty. There is no running water on the *Sarah*. Please give me strength and the wisdom to do what's right to get RNI back on the air.

A personal note: I am alone. I do not know if my wife Sarah and I will ever get back together again. J.P. says, "When it's over, it's over." But he has been wrong before. Should I see another? Strange to work it out on paper. Anyway, I don't know if I can take the strain and pain of going through a brand-new relationship. I met a gal named Sarah; yes, they run in threes... all driving me crazy.

(I met a reporter whose name was Sarah. She came on the ship a couple of times. Being that I was so lonely, I'm sure I must have asked her out. She seemed nice, but had some basic problems —

not liking most people and believing that love is an illusion. Love is probably the only real thing that two people have between each other. What can replace the feeling of two people holding hands or giving each other a hug? What would you trade for that? Love, compassion, and understanding are the things that make us human.)

I could pursue this, but I feel it could take a lot out of me, and maybe her. And I still feel loyal to my wife, Sarah — even though she has not been loyal to me. I do not know what to do. For 15 years, I have basically been turned off to anybody but my wife. We loved each other very much, and I could never be interested in anybody else. There wasn't even a thought of it. My heart is in a state of confusion. What shall I do? I really don't know.

RNI is off the air, my wife and I may be finished, and I am interested in another person named Sarah... why me? Something somewhere must be getting a laugh out of all this mess. But that's life. There are no guarantees of fairness. It's just that I am so lonely.

I told Frank that I could ride up on the ship for a while, but he was going to have to let me off somewhere in Cape Cod. Then I'd catch a taxi, catch a bus, and hopefully have enough time to get down to New York City for the filming of the MTV segments. In addition to towing the *Sarah*, Frank was pulling another small boat alongside. But early that morning, when we were just out of Buzzard's Bay, it decided to sink. So Frank had to stop the *Munzer*, stop the *Sarah*, and drop the anchor chain to keep it in position while he tried to salvage it. So I got on the *Munzer* and Frank dropped me off at a military installation somewhere in the Cape Cod area.

It was early Saturday morning, and I had to get down to the bowels of New York City by 2 or 3 PM. I walked to the closest town and was searching for a bus station so I could get to Providence, Rhode Island, and rent a car to drive to New York. I couldn't locate one, so I went to the local police station. I walked up to the dispatcher's desk and said, "Officer, I need the directions to the nearest bus station."

He said, "It's down the road about a quarter of a mile, take a right and you'll be right there."

I thanked him and started to leave. But then he started staring at me. I guess was a little grungy and dirty looking from working on the ship, and for some reason, that seemed to bother him.

Then he said, "Before you leave, how about showing me some identification?" I was shocked. I went to the police station to try and get some help, and this guy wanted to interrogate me.

I said, "Gee, Officer, why do I have to do that? I haven't done anything wrong."

Then he got real nasty. He said, "If you don't show me some identification right now, I'm gonna arrest you and throw you in jail."

I was in one of those moods where anything could have happened. But I guess the good Lord or logic prevailed, and I whipped out my driver's license and handed it to him. I had to sit there for five or ten minutes while he ran the stupid thing through his computer. Then he came out, handed it back to me and said, "Fine, have a great day," and that was it. But the idea of walking up to the friendly policeman on the corner for directions sure got shot to hell in that particular incident.

I found the station and got on the bus to Providence. Then, after I arrived, I had the great hassle of trying to find a place that would rent me a car on a one-way situation. I had no luck with that, but I rented a car anyway, got on Route 95, and drove like a bat out of hell to the MTV studio in Manhattan. Somewhere en route, I managed to duck into a bathroom, clean myself up, and make a change of clothes. When I got to the studio, the whole gang was waiting. It was great to see everybody again.

The studio session took a couple of hours. We took turns doing intros for their music videos. They taped us saying things like, "Hi, this is the RNI crew and now we're going to see..." We did a whole bunch of those in succession. Everybody was really nice to us, and MTV was a cool place to be. After it was over, I hung out with the gang for a while.

Around that time, some friction was starting to develop between Randal, John, and Perry. While the ship was being outfitted, everyone got along fine. There were no problems. But for some reason, the WNYG gig was creating a lot of tension between those guys. John and Perry had their own ideas on how they wanted to do their shows, and Randal had some different views on how he wanted to see it happen. I stayed in the background as much as possible and did my best to act as an advisor/consultant to both parties.

During this time, Randal called me up quite a bit. He was upset because John and Perry were not listening to his advice as operations manager. In the meantime, John and Perry decided to start their own production company because of all the media attention we were getting. They wanted to separate from RNI and do their own production thing. The guys still went out on the weekends to do their program on WNYG, but sadly, our organization had started to deteriorate.

I was disheartened by it all. We were in conflict with the government because they had destroyed our station and wouldn't let us broadcast. That was the main issue — or so I thought. I felt that the best way to deal with this was to stay together and fight it as a group. But I could see that was not going to happen. I did what I could to settle them down. Then I went up to Maine to get some peace.

By August 6, I was back in Monticello, the ship was safely back in Boston, and I had done many interviews — WNBC, WBAI, *Newsweek*, the *Voice*, *Rolling Stone* magazine, etc. The preceding Tuesday, my lawyer and I had met with Isaac, the owner of the Hard Rock Cafe in New York City.

(After the government tore up our station, Isaac had a whole bunch of flowers and plants sent to my lawyer's office with good wishes for RNI. Then Isaac contacted Dudley about possibly financing a relaunch of the ship, which was pretty exciting.)

I stayed in Monticello for some rest and relaxation. The quiet and lack of cars is a blessing. I do miss the solitude of this place when I am gone. The *M/V Sarah* was fine in Boston. A bit plundered and damaged, but still afloat. It would be possible to make her operational again, but I calculated that the cost would be at least $10,000.

I couldn't believe that I was off the air again. The government had been very cruel about all this. All we wanted to do was serve the public and raise some awareness. I guess those are two things the government fears.

In early August, I met with my lawyer, Dudley Gaffin. We spoke of the case, how the government carries out its various illegal acts, and what we were going to do:

1. Clear Ivan and myself of charges

2. Sue the government and FCC for damages

3. Get RNI back on the air.

A few days later, I spoke with the owner of the Hard Rock Cafe about helping RNI. He was very interested in our plight and gave me a check for $5,000, plus a letter of commitment to help us along. I wrote in my journal: "I hope we can get more support from Isaac and The Hard Rock to get RNI back on the air. Finally, someone who cares."

Isaac was a nice guy. He had gone through a series of hard knocks to build his Hard Rock Cafe. He was interested in helping us because he felt our fight to gain access to the airwaves was very similar to his struggle to gain recognition for his type of rock and roll theme cafe. He felt that we were doing a similar thing, only with radio, to what he was trying to achieve with his restaurants.

Our lawyer, Dudley Gaffin, was just great with us. He seemed to really care about our well-being, both legally and humanely. I hoped we could raise some money to help offset his expenses. (He really put out for us, and didn't charge us a dime. It just goes to show you that there are some good-hearted people on the earth.)

On August 24, I learned that there would be no hearing. The prosecution would go for a grand jury indictment on the charges outlined in the complaint. We would have to fight it out with a full jury trial. Ivan would have to decide for himself whether or not to reject their offer of deferred prosecution and be willing to assume the risk of heavy fines and/or jail.

The government tried to get us to accept a deal where they would defer prosecution if we agreed to plead no contest. Ivan and I decided not to go that route. It was like admitting defeat, or saying that we did something wrong.

I believed in what I was doing. RNI was in the right. To give in would set a dangerous precedent that the government could go anywhere they chose on the planet and pirate vessels — board, arrest, loot, and pillage. Also, there was a port order against the M/V Sarah preventing her from leaving Boston until some conditions of insurance and safety were met. It was just another way to hassle and keep RNI off the air.

In late August, just 3½ hours before the hearing was designated to start, the government people called Dudley's office and said, "We've decided not to press any charges." I assume that they sat down and realized that it would be difficult to prove anything against us — so they dropped all the charges and the matter was closed. So we won the first round.

Quite honestly, the government didn't have a case — and I'm sure they knew it. We didn't violate any FCC rules and regulations and I don't know where the hell they dug up the charges they were trying to bring against us — some obscure law about conspiring to impede the functioning of the United States government, which can carry a $250,000 fine and five years in jail.

At last, we could try to get RNI back on the air. We needed to raise some money — maybe the Hard Rock would help.

Anyway, Ivan and I were off the hook for a while.

Chapter Nine
Getting the Ship Back
On the Air

In the late summer and fall of 1987, I was doing my best to get Radio New York International back on the air again. The main problem was that everything associated with ships is expensive, and there was just no money available to make all the repairs that were necessary to get the ship out of Boston and back to her Long Island anchorage. But with the little money I had, I did what I could.

Dudley and I discussed the possibility of filing charges against the government for the damage they did to the radio station on the ship. We also talked about initiating a procedure to restrain the government from coming out and tearing up the equipment or arresting anyone when we went back on the air again.

Outside of all the conflicts I was having with the FCC, good things were happening in my personal life. Elayne Star and I started going out on August 25th of that year, and it wasn't long until we were very much in love. Even then, I could really see us having a lot of good times and sharing our lives with one another. By that time, I realized that my life with Sarah was over. The final choices had been made. She decided that she wanted to stay with her girlfriend, Rose, and it was time that I moved on with my life as well.

In mid-September, Dudley and I met with some people from the Hard Rock Cafe and they agreed to fund the refitting of the *M/V Sarah* for a return voyage off of Long Island. About $200,000 was involved in the deal, which Dudley set forth on paper. It was really hard to believe that they were going to provide funds to get RNI back on the air. If it worked and RNI was able to stay on the air, they promised to provide the backing to bring out a million-dollar vessel and do it really big. So I went out to the ship and started to piece the main studio back together. It felt strange to have to rewire the console, turntables, and tape recorders all over again. But my hands just seemed to know what to do, and within a few days it was all finished.

By October, I was back in Yonkers, hanging around and seeing Elayne. I hadn't received any funds from the Hard Rock Cafe beyond the $5,000 Isaac initially gave us. It was very frustrating. I was living off my supplemental income and what was left of the RNI account. I was $30,000 in debt for that rusty old ship. Winter was coming, and we needed to move ahead soon.

There was some good news though — Mike Barrington from England had agreed to come over and engineer aboard the *Sarah* — or at least look the ship over.

That fall, things were moving very slowly, but I did my best to keep the morale up among the crew. Josh Gaffin (Dudley Gaffin's son) put together a fund-raiser concert that was scheduled to take place at the end of October. It helped to keep RNI alive, as well as in the news. I wanted to get the ship out as soon as possible. But with no money in the RNI account, I knew it would be very difficult indeed. She was without a prime generator, but still afloat.

Around then, there was some fighting involved with RNI's takeover programs at WNYG. Randal was at odds with John Calabro, and all of them were feuding with the owner of the station. For some reason, John and Randal's personalities just kept clashing. I guess that's just the nature of the business, but the whole situation was breaking my heart. (The folks that work with offshore radio stations in England told me that the same thing happens over there. People come in and fight with each other and fight with the management and fight with the owners — then they usually end up leaving.)

On October 12, with no funds coming in from the Hard Rock Cafe or anywhere else, I went back to Monticello, my fortress of solitude. Dudley said it's unlikely that the Coast Guard would ever let the *M/V Sarah* leave Boston as a radio ship. He thought that the only way to do it was to get another ship to come in from a foreign port. He also felt that I should continue to pursue the idea of obtaining a license to build a shortwave station in Monticello. Meanwhile, I called Peter and arranged to have the ship painted before winter set in. The job cost me $400.

In late October, David and Isaac from the Hard Rock Cafe came to look over the ship. They were both impressed, and said that they wanted to go ahead and work together. Then they threw a proposal at me saying, that for $25,000 down, they wanted the ship plus all movie, book, and commercial rights. Then in six months, they wanted the option to pay an additional $25,000 to secure the ship — as well as all promotional rights. In other words, they wanted to own all of RNI lock, stock and barrel. I was stunned. What happened to the REAL deal of seven weeks ago, which was basically refinancing the whole operation and getting it back out there and on the air?

Dudley and I worked for several hours on a counter-proposal, which we submitted to their lawyer. We had no word on that for weeks, and I wasn't able to contact anyone and find out what happened. In the beginning, the Hard Rock Cafe people seemed very promising. But as we tried to negotiate a deal, the money never materialized. The Hard Rock Cafe was really into it, but as their lawyers pondered the many legal questions that offshore radio poses, I think they got cold feet. It would have been so much better if they had come to us and said, "This seems to be too much of a borderline investment issue for us, and we can't do it."

I would have said, "Fine, thank you for your support, thank you for your help, thank you for the $5,000." Then we could have started working on other options. But as it was, they led us on for months and wasted a lot of my personal time in the process. By then, I realized that if RNI was ever going to be relaunched, I would probably have to figure out a way to finance it myself. I talked to John Ford from New Jersey, and he suggested that we raise some money by renting out the ship as a promotional billboard.

Meanwhile, the crew was fighting among themselves. Randal was fighting with everyone, and everyone was fighting with Randal. If only all that energy had been directed towards rebuilding RNI, doing programming, fundraising, selling ads, and things like that! We would have been in much better shape.

In late November, I went to a meeting with the RNI staff at J.P.'s apartment. After about two and a half hours, we worked out the problems between Randal, Josh, John, and Perry for the time being. Then we talked about raising funds and making money through promoting the RNI story.

The M/V Sarah *at anchor, September 1988.*

By then, I was beginning to feel that we should bring the ship back out to its Long Island anchorage and just let it sit there for a while. That way, all the forces against it and all the media people and friends of the station would have a chance to see it and get used to it being there before any transmitters were switched on. That way, we could get media coverage, announce our on-the-air date, and watch what happened. (We didn't end up doing it that way, but it was a thought.)

I met with the Coast Guard and discussed what the owners of the *Sarah* must do to get her out again. The Coast Guard was friendly. They just cared too much about all their procedural things. We had to do all kinds of stuff before we could take the *M/V Sarah* out to sea because she weighed over a certain tonnage.

Around the middle of December, I went on a cruise to Central America with my father. Boy, did I need it. I spent one and a half years and $76,000 on RNI, and had nothing to show for it. (Some people think I spent around $200,000, but it wasn't anywhere close to that amount. That's what I'm good at. I can take very small amounts of money and build radio stations — AM, FM, shortwave — whatever.) Every time I take a cruise, I always think of different ways of making money. But it always falls back to radio engineering. Maybe I'm the last of a breed.

Just before Christmas, *The New York Daily News* called me for an interview and *Popular Communications* asked me to write an article for their magazine. Other than that, not much was happening. I knew that I should find some kind of part-time work to keep me busy. But whether it would be in Maine, Boston, New York, or Florida was still a mystery to me.

Whether or not I could find work was also a big question. Unfortunately, it seemed that engineering is getting real low on the list of useful jobs. I really missed being on the air. And even after all I had been through, there was still no signal for RNI. I wrote letters to everyone from Ronald Reagan to Donald Trump trying to raise a buck for RNI. I even thought about writing to the FCC and asking if they would give me KPF-941 back. A waste of time, no doubt, but at least I could say I tried. As it was, I had nothing but a frozen ship and a cold winter ahead.

1988 was a brand new year. What would it bring? Hopefully, no more trouble. 1987 was quite a gas. Losing this, gaining that. WREM Radio signed on the air Christmas Eve as a religious station — Love 710, under the direction of Donny Flewelling. I prayed that I did the right thing.

Ivan and I froze our butts off in late January while having our photos taken for *Rolling Stone* magazine, which named us the Critics' Choice for best radio station in 1987. (Can you believe it took them 2½ hours to take a photograph?)

That winter, most of the pirates in the New York area that went off the air for a while to concentrate their efforts on producing programs for RNI started broadcasting again. And as always, I advised them to be careful about the FCC. Then, in early February, John, Perry, and Pete told me that they were going to resign from RNI because they didn't like Randal. After that, there were only myself, J.P., Elayne, Randal, Josh, and John Ford to run RNI.

Meanwhile, I was trying to nickel-and-dime my way back on the air. I went to the farm for a while and did some work with RNI Surplus. I sold a few items and raised some cash for RNI's return to the airwaves. The Hard Rock Cafe still hadn't come through, but Pop forwarded me $5,000 to get me started. On March 27, I met with Julian, the representative from the Honduran Maritime Bureau, and began the long, expensive road to permanent registration.

By mid-April, Honduras was proving to be a real pain. They wanted $4,000 for registration — if they decided to accept it. They said that the *M/V Sarah* was not seaworthy in its present shape. "A dead ship," the surveys said. But just batten down the hatches, and she'd be watertight and storm-ready. This project chipped away at me every day — problems, problems, a thousand problems. No breaks, just one tough mess. But there are always alternatives to any troublesome situation. And I knew that I would find them.

My friend from Florida, Noah, bought a share of stock in RNI for $2,500. Finally, someone who backed their convictions with cash! Furthermore, she was going to see if her grandmother would invest more heavily for her. Then we would have enough to do what we need to do.

A few days later, I went back to the ship and did some work. I spliced the FM transmission line, completing the repairs of the damage caused by the FCC. Then I drove up to Monticello. I was trying like the dickens to find a way to haul that 400-ton floating transmitter site out of Boston Harbor and down to its anchorage off of Long Island. I knew it wouldn't be easy — but nothing is.

In the early weeks of May, I went to New York City to visit with Elayne and have some meetings with Dudley. I told him how things were going with the ship. Dudley said he talked to a fellow in Los Angeles who said that he was interested in promoting RNI as a movie and wanted to work up some kind of a deal. Little money was offered up front, but bigger bucks were in store if the picture was a success.

I also spoke to Uncle Jim — the guy that generated all the investment to re-launch Radio Caroline in 1983 and bring out Laser Radio in 1984. He also sounded like he might be interested in financing the ship. (Of course, nothing came of any of that, but it did give me hope for a while.) Then I went back to Boston to work on the ship and weld in the generator we were finally able to obtain.

In early May, I drove down to Uncle Jim's place in Smithville, New Jersey. We spent the day talking ships and offshore radio. He had big plans — he wanted to equip four big oil tankers with radio stations and link them all by satellite.

He said he wanted to back our venture. So far, it all sounded good. But when it comes down to writing out the check, we would have to wait and see. (I was always realistic about that. I know that most people are 99% talk, and 1% money. But depending on who they are, that 1% money can be a lot.)

A contact of Uncle Jim's said that he could get the ship registered as an non-self-propelled barge. Anything to get away from the Honduran registry would be a big help.

In mid-May, I was hired at ABC for vacation relief work.[1] May 17 was my first day. It was not my favorite situation, but the people were nice and the pay was good. It felt strange to be working in New York City, but at least I could spend more time with Elayne. I was still trying to get the ship registered, and I was thinking about hiring an attorney in Boston to help me work around all the Coast Guard problems and get the ship out.

Uncle Jim told me that he was still going to back the venture — but when it was time for him to produce some money, he proved himself to be a lot of hot air. When I asked him for the $2,700 I needed for the registry, he stalled and said he would have it when his loan came through. It seemed that everyone was full of crap when it came to financing this project.

In June, I tried to acquire an Antiguan registry. But I couldn't seem to do it without proper surveys. Being that the ship was old and in tired shape, that was a problem. We did our best, but we just didn't know what the results would be.

By July, I was working at ABC, weekends and all. The ship had been painted black with her name in white lettering, all in preparation for her return to the high seas.

---

[1] How I got the job at ABC is as follows: I took an ad out in *Radio World* magazine, which is a trade publication for the radio industry. The ad read something to the effect of, "I built the radio station and radio ship Radio New York International. What can I do for you?" Then I got a call from a fellow named Karl Zuk who writes articles for *Monitoring Times*. Karl had read my ad in *Radio World* and asked me if I was interested in doing vacation relief work at ABC television. Of course, I said, "Sure, why not?" So he gave me a bunch of numbers to call.

The first thing I did was call J.P. and say, "Hey J.P., they're looking for people to do vacation work at ABC." They were looking for people who knew what audio circuits were — balanced lines, impedance matching, amplification, and so on. And luckily, that was right up our alley. We went in for a couple of interviews and were both hired to work there.

I went to Boston a few days later and spoke with Frank. We still had some Coast Guard problems to take care of. They just loved the new black paint job. But unfortunately, all the notoriety presented a few problems. John spent the week on the ship painting and helping out where he could.

Frank tried to take the *M/V Sarah* out early one Sunday morning in July, but the Coast Guard caught on. They came out with their boats and stopped him just short of clearing the harbor. So that was the end of that idea.

I met with the Coast Guard, and they were still holding firm on their port order. Many of the items they wanted were almost impossible to obtain, and I was starting to feel that this was a no-win situation. I was started to consider the idea of selling the *Sarah* and getting another ship for the project..

(The Coast Guard was enforcing regulations that should only apply to fishing vessels and vessels that move cargo. The *M/V Sarah* was refitted to be a floating barge with a radio station on it. But as always, bureaucrats were good at being bureaucrats.)

Around the end of July, I was on the phone battling it out with the Coast Guard and insurance companies. I had to disconnect and seal up the fuel lines, tank lines, and tanks that weren't even in use. It was a totally useless job, but it had to be done so the ship would be exempt from all those crazy oil-pollution regulations.

The damned Boston Coast Guard also refused to let my ship out unless it had liability insurance. I checked around and found out that it might cost me $3,500 just to cover a one-hour journey from Boston into international waters. Such nonsense!

I was really getting tired of being beaten down by the government. I wanted to do some real radio broadcasting to serve the New York area. Why was that such a problem? All this idiotic crap was draining me, but I vowed not to give up. No matter what, I would find some way to persevere.

By mid-August, I was still seeking insurance for the ship. My agent in Boston got a "yes" from Lloyd's of London, but they wanted a survey to make sure the ship wouldn't sink. All of this

took a lot of time. Too much time. I did finally register the ship —
the *M/V Sarah* was registered in the state of Maine. That I hoped
would make the Coast Guard very happy.

(I went to my town office in Maine and found out that I could
register the *M/V Sarah* as a State of Maine Vessel for $4. And then
when the ship was out in international territory, I could cancel the
Maine registration and plop on whatever foreign registration I could
obtain for it.)

I got a call from Dudley. He said that the government was trying
to get a restraining order against me, J.P., Randal, Ivan, and RNI.
From what I understood, it even extended to land operations. Since
this was not a criminal matter, I was told that we had many rights
and various courses of action.

(The government was trying to get a permanent injunction
against us to keep us from broadcasting from anything from a
bathtub to the planet Mars. It was purely prior restraint and
harassment. It was a civil, not a criminal procedure. But still, we
had to address it.)

It looked a bit doubtful if I would ever get the *M/V Sarah* out of
Boston Harbor. Getting insurance was a long process. It's crazy to
be forced to spend thousands of dollars for insurance on a 60-
minute jaunt out of the harbor. I considered selling the *Sarah*, re-
grouping, obtaining a vessel under 200 tons, and trying it again. (If
a vessel is under 200 tons registered weight, a lot of insurance
requirements and pollution control regulations don't apply.)

I felt that I had made some key mistakes so far in the venture:

1. I should have bought a ship in better shape — and under 200
   tons.
2. I should have been better prepared when we went out the
   preceding year. Going on the air as fast as we did may have
   been a mistake.
3. Letting the Coast Guard and FCC on the ship so early may
   have been a mistake.
4. The ship should have remained out at anchorage. It would have
   been better not to have brought the *Sarah* back to Boston —

especially with all the problems that arose with the Coast
Guard in trying to get her out again.

5. Giving up so easy last summer was a weakness on my part. We
should have stayed out there and fought for our rights.

6. Of course, not having enough working capital has been the
Achilles heel of this whole project. No money = no nothing.

(Of course, it's debatable how much of a fight Ivan and I could
have put up on that fateful day when the government boarded the
ship and took it over by force. Looking back, I think I did the right
thing in not resisting, because, as it turned out, nobody got hurt.
That's more important than anything. If anybody had been injured,
it would have permanently soured the whole project, as far as I'm
concerned.)

After an entire day of fighting with the Coast Guard to prove that
the *M/V Sarah* had the proper insurance, they finally released the
ship at 2:30 PM on September 8. First, they lost the telex from
London, then another was sent. It turned out to be the wrong one.
My agent had London reword it to their liking. And at long last, the
Coast Guard finally decided to accept it. Frank was set to leave as
soon as possible. And my first crew, Reggie and Josh, arrived at the
Sarah in Boston.

I met with Jeremiah, and he agreed to represent us before the
courts in our big civil matter with the FCC. He said that we would
need to show that they had no jurisdiction over what we were doing.
And he wanted to go even farther and challenge the entire FCC
policy not to license people who want to broadcast on unused
channels. (Jeremiah is the same fellow that J.P. and I had met with
some 17 years before after the bust of WXMN and WSEX.)

By midnight of September 10, *M/V Sarah* had finally arrived at
its Long Island anchorage. Frank left Boston on September 8 with
Reggie and Josh on board. I arrived on Long Island Sunday to go
out and meet the ship. Upon my arrival, I found little food or fuel.
(That's how Frank got the nickname "Half-Tank Frank." We
bought a full load of fuel, but somehow we only got half of it. That
night, I spent a long time cleaning fuel. Apparently Frank had

pumped some fuel on board and it got mixed in with water. So we had to pump the fuel into the day tanks and wait for it to separate, then separate the water out and clean the water separators. It took a long time. But I guess that's just the way it is in the offshore radio business.)

In late September, the ACLU decided to take our case. Finally, a break! They determined that we had First Amendment rights which were being violated. Jeremiah requested a change of venue from Boston to New York, and filed a Freedom of Information Act request on behalf of me and the *M/V Sarah*.

On September 21, I went to England to attend "Communicate 88," the first Free Radio Convention in Blackpool. I had a great time. I met lots of people and was interviewed, spoken to, asked to sign autographs, and more. There was a genuine free-radio spirit there, along with a strong unity. I only wish it was as together in the United States.

I must say that the various free-radio support organizations had their act in gear. There were dozens of pictures, books, tapes, T-shirts and the like for sale. Even medium-wave loop antennas were available.

But when I returned from my trip to England, I discovered that we almost had an abandoned ship situation on the *M/V Sarah*. The ship ran out of fuel, and there was nothing to run the generator, which made it pretty rough for the guys.

(The ship was out at anchor, and the plan was to just leave it there for a while. I was curious to see what would happen when they started broadcasting, but I didn't dare turn it on the air until we reflagged it with a foreign registry.)

I knew I had to get the ship back on the air somehow. So I decided to talk to the Bates brothers of Sealand about registering the *Sarah* in their independent state.

In the 1970s, Michael and Roy Bates took over an old World War II offshore anti-aircraft platform in the North Sea. They declared it sovereign and went through all the necessary paperwork — they even issued currency, stamps, and whatever. They made it into a small nation, which they called "The Principality of Sealand," and posted people out there to defend it. I know it sounds crazy, but that's what you do in the offshore business. As long as it was legal and it made sense, we would do it.

Michael Bates and I worked out a deal whereby if they gave us registry for the ship, we agreed to give them air time they could use to promote their country, products, or whatever they wanted to sell. We decided to set up a London company called Atlantic Radio Communications Limited. And in early October, the *M/V Sarah* was duly registered to the Principality of Sealand.

Vincent agreed to help me set up a British company to own and run Radio New York International. That would alleviate a lot of responsibilities on my part and make everything more legal and proper. So, as soon as I got back home, I had the papers drawn up and the whole bit. We also talked about going stateless on the vessel, but I didn't think that would be a good idea.

Again, we felt just as we did in 1987. It was unfinished business. We knew we weren't violating any rules or regulations of the FCC or any other government agency. We just wanted to bring the ship out, go on the air, broadcast rock and roll, and alternative programming, and be a free radio voice. That's all we wanted to do. We didn't want to hurt anybody, and we didn't want to jeopardize the whole radio industry. We just wanted access to the airwaves.

Chapter Ten
The Station Goes On,
And The Feds Crack Down

On October 14, 1988, at 9 PM, RNI returned to the air. (I was not on board, by the way). Captain Josh put out the first words. Three hours of music and IDs followed. I listened to the signal in Yonkers and at Elayne's apartment in Manhattan. J.P. listened in his mom's basement. Even Karl Zuk picked up the signal down in the bowels of ABC. It felt great to hear Radio New York International on the air again.

Vincent said that if we could last a week, we were home free.

The second night we went on the air, we only broadcast on 1620, just above the top of the AM band. The power level was a little over a kilowatt. I didn't want to push the full 5,000 watts — at least until we saw what happened. We used a sloping T antenna. And from the vantage point of a ship in saltwater, it really put an excellent signal all over the New York area.

On October 17, Elayne called and said that the court in Boston had issued a temporary restraining order against me, J.P., and Randal — as well as the *M/V Sarah*. The restraining order had yet to be delivered to the ship. Elayne notified Captain Josh, and he was ready.

At 9 PM, RNI signed on for the fourth broadcast. At around 10:30 PM, a 41-foot Coast Guard boat arrived at the ship. For about thirty minutes, Captain Josh and the Coast Guard had an exchange of words. Josh denied them permission to board. The Coast Guard said they need not board the ship to issue the order. And they threatened that if RNI didn't go off the air, the court would allow the FCC to board and dismantle the station. Needing more time to consider the information, Captain Josh signed off around 11 PM. The Coast Guard backed off and circled the *Sarah* for a while.

So RNI was off the air again. The main concern here was the ship and crew. We did not want the crew hurt or the *Sarah* trashed. Clearly, they had no jurisdiction. Jeremiah said the government doesn't care about rights or the law. They do whatever they want. They would wreck the vessel, arrest everyone, and sort it out in court later.

I was getting confusing advice from all sides. But everyone agreed on one thing — we should go to the hearing on October 21 and see what happened.

J.P., Jeremiah, and I went to the hearing in Boston together.

First, the government presented their side, which was the same old arguments about why the FCC must regulate the airwaves, how the bands are limited resources, and how they have the authority to regulate any radio signals entering the United States.

Then it was our turn. Jeremiah did a great job. He argued for the First Amendment and our right to broadcast on unused frequencies, such as 1620. He stated that the FCC was silencing our voices. Everything that should have been said in August of 1971 was said, and more.

The judge seemed interested. After listening to both sides present their cases, he commented on how important the issues were. In six weeks, we would know his decision — that is, if he didn't let it get referred to a higher court. (That's the way it works in civil law. You present a case, and then you have to sit around and wait for the judges to come up with a decision.)

We, as well as the ship, were still under the government restraining order. To go on the air would surely result in a seizure of the vessel, and we would probably be arrested for contempt of court. A busted-up ship and our imprisonment would not do our situation much good. So with great sadness, I had to concede that the logical thing to do at this point was to try and lease the *M/V*

Sarah for use outside the close vicinity of the United States — then see the case through to its logical conclusion.

On December 20, the judge finally arrived at a decision. Guess what — we lost. He ruled on incorrect facts. He assumed that the *M/V Sarah* was still registered in Maine, which was not true. In any case, it was as expected. The government will lie, distort the facts, and make up any rules they like. Jeremiah would appeal and try to have the restraining order lifted, but I thought it would be a futile attempt, at best.

The judge ruled that the United States government has jurisdiction over all radio transmissions that not only go out of the United States, but come in as well. So basically he was saying that the government has the right to control and regulate Radio Canada International, Radio Moscow, Radio Japan International, Radio Australia, The Voice of Taiwan — all the international shortwave broadcasters — plus all the satellite transmissions you can pick up in this country.

That is the true interpretation of the judge's ridiculous ruling — which is preposterous and has no foundation in international law. But that's what can happen when you go into civil court. Sometimes you get a judge who doesn't realize the technicalities and nuances of broadcasting.

The FCC went in there and sold the judge a bill of goods, and that was it. They came into the courtroom with sheets of paper that listed hundreds and hundreds of possible stations that were allocated to use the frequencies of 1620 and above. J.P. and I were totally stunned because 1620 had been a dead frequency for years. And the proof of that is that those frequencies are now being allocated by the FCC as an extension of the AM broadcast band.

I still wanted to fight — but I didn't want to commit suicide. It was really a painful time for all of us at RNI. Even after the judge's horrible decision, I knew that Josh and Reggie were ready to go on the air at a moment's notice. If I had called them up on the cellular, and said, "Just do it," they would have. But I knew what the U.S. government would do. They would go out there to slash and burn and go crazy issuing more injunctions and contempt-of-court warrants. It was just getting insane.

At this time, we had very little money left. God bless Jeremiah and his associates at the ACLU. They really put a lot of time in, and they did the best that they could. But we were all pushing up against a brick wall.

Then on December 30, after 200 years of precedence, Reagan gave an executive order extending the territorial limits outside the USA to 12 miles.

Now let's think about it — here is my ship, sitting out at anchor since September of 1988. We are having this battle with the FCC and the government over getting the ship on the air and our main point is that it's in international waters, which is free territory. And lame duck president Ronald Reagan in one of his last acts in office signs an order extending the territorial limits of the United States of America from three miles, which it has been for the last 200 and some odd years, to 12 miles.

When I heard the news, I was stunned. Why at this very time in the midst of a legal argument over the broadcasting rights of one ship in international waters does the president choose by executive order to extend the territorial limits of the United States?

So the *M/V Sarah* was now in U.S. waters. Nice, wasn't it? The only good thing was that since the ship was in U.S. territory, I didn't have to keep a person on board all the time. (If a ship is on the high seas and unmanned, the right of salvage takes place, and anyone can come and claim the boat. But once it is inside of the territorial limits, its like keeping a ship at anchor in port — you're protected under personal property law.)

The ship was now available to anyone who could come and move her somewhere on earth where she could broadcast in legal peace. I was not giving up, but I had to be practical. This constant outlay of cash and nothing coming back for two years had been a real drain. Now I just wanted to see the ship used somewhere.

By January, 1989, everybody was exhausted. I was out of money and so was J.P. The ship couldn't broadcast because the restraining order was still on, and we were going to be in appeal for about six months. When Reagan extended the territorial limits, that was the last straw. After that, keeping the ship in international waters meant moving her out an extra 10 miles or so — plus buying a longer anchor chain. I was not in the position to do that. So I took a few things off the ship, locked it up, and told Josh he could come off. And the *Sarah* became a 400-ton buoy, sitting out there with a light on it.

In early January, I filed a letter with my pals at the FCC requesting to have my application for an international land-based shortwave station granted. I realized that I was not going to be able to do anything with the *M/V Sarah*, and I still had an application for a shortwave station pending before the U.S. government. So I decided to type them a letter and see if I could reactivate that application. I noted in my journal, "Once the FCC stops laughing, the long battle to regain my rights to be a suitable licensee will begin."

For the rest of the winter of 1989, the ship sat at anchor, and I was at the farm in Monticello for most of the time. I came down every once and a while to visit with Elayne, but I spent most of the winter at the farm trying to get something done with the *Sarah*. I was also doing some general engineering for a few radio stations in Maine.

In March of 1989, Tim Sexton came through with some money and a signed contract. He contracted with me to buy an option to do a movie on the RNI story, which would have featured the ship, and he wanted to make it into a musical presentation. Nothing ever came of it, but I did get some pocket money, which certainly helped things along at this point.

Also, John X had been calling me up literally every other day with a different idea. The plan that week was to bring the ship to Europe. We were going to put somebody on the ship and get it ready to go over at the end of June. John X came across with some money, so we put a cash deposit on three shortwave transmitters from good old Fair Radio Sales Company. Of course, our plans never materialized.

In April, I went to New York City, and went back to ABC for another season of vacation relief work. At that time there was much talk of radio ships going on in Europe. The *Communicator* had moved, and another radio ship was supposed to appear on the European scene (but never really did). Uncle Jim spoke to me about the two former U.S. surveillance ships that he was trying to get hold of and outfit as radio boats. And the *M/V Sarah* was still at anchor at Long Island.

In mid-May, the ship was brought back to the Bang Ship Yard in Boston. Frank reported the *M/V Sarah* to be fine and dry. The antenna structure was A-OK. Both of us were amazed at how well the ship took all of the terrible storms and generally being knocked about. I don't want to brag, but I can build them. Now it was time to make her ready for the voyage to Europe. John X's partner was due in New York the following week.

Late June arrived, and not much was happening with the ship. The vessel was still berthed in Boston. John X and his partner said they planned to paint and repower her, but after a lot of talk with Peter, Frank, and me, no money changed hands. (That was normally the rule with this thing — there was always a lot of talk, but never any money or action. In the offshore radio business, that's the way it is.)

On July 29, 1989, we had a free radio convention. (Occasionally, J.P. and I held free radio conventions. This was the third or fourth one that we had held in the past 20-25 years.) Actually, it was just a get-together of radio enthusiasts. I was there, as were Randal, the people from WHOT, and some other people who just heard about it. It was fun. We sat around and talked about radio, getting on the air, and what needed to be done. It was a nice, warm pleasant summer day.

During the spring and most of the summer, the ship sat in Boston and John X and his partner were running around saying that they were going to do this and do that, but nothing really ever happened with it.

J.P. and I traveled to Boston to attend the appeals hearing in early August. Jeremiah said he was satisfied with our position. He hoped we'd win and get back on the air. Jeremiah presented an oral argument. Basically, he argued for our First Amendment rights, and that was it. It was another one of those anticlimactic things where he stood up in front of the tribunal of judges and said, "I feel my clients should be allowed to go on the air because they have a right to use these unused frequencies, and the First Amendment guarantees this and etc..." Of course, many months later when they came back with a decision, it was thumbs down again.

On August 19, I was having lunch at Elayne's apartment when I got a call from one of the people of the Caroline organization informing me that Radio Caroline had been raided. I was shocked, I really was. The Dutch government just took a tug boat out there and raided the ship. They literally pirated the vessel. They had no documents or anything... they just came on board and strong-armed their way onto the ship. There were only about two or three people on board at the time.

The government agents used sledge hammers to break up one of the transmitters. They took all the records and studio equipment off the ship and left. It created quite a furor. But that's what happens when governments think they own the bloody blasted airwaves. If anybody dares to do anything, even if it's contributing to the whole of society, they'll destroy it. This wasn't the first time in offshore history that kind of thing had happened, but it was very surprising that it would occur in international territory near an advanced, seemingly tolerant nation such as England.

After my initial shock, life went on as usual. By summer's end, J.P. and I were regular employees at ABC. A full time wage-slave job.

I had a love/hate relationship with the job at ABC when I was there. The money was good, but it was a typical job where your time is not your own. I had to be there at a certain time, and I had to leave at a certain time. I figured that I'd work at ABC for five years and get the experience, then move on to something else.

Around this time, I closed the deal on a house in Yonkers. I was beginning to get the stuff and furniture that I had stored in Monticello, Maine, together so I could bring it down and move it into my 90-year-old home in Yonkers, New York. It was a damn nice house. It was about 2,400 square feet, and on a small lot, but it was in a very, very old section of town.

I was fairly satisfied with my purchase, but Elayne hated it. She thought I was crazy to buy anything in Yonkers, because it was turning into just another town of urban decay. But like anything else, even in the middle of Brooklyn, some areas are still kind of nice. They have beautiful homes that are still kept up by their owners. And the Park Hill area was very similar to that. The price was a little ridiculous — but at the time, it was relatively inexpensive.

In early November 1989, Johnny Lightning's station, WJPL, was busted by the FCC. The government took all of their transmitting gear. This seemed to be their new tactic. 1989 was the year that the government went into their new *modus operandi* of arresting equipment. To me, it seemed to be illegal search and seizure. I don't understand how they can arrest equipment when it didn't do anything. It's like arresting a gun that you shoot somebody with. What does that mean? Anyway, the bust of WJPL and the confiscation of all of Johnny Lightning's equipment was the beginning of the FCC's new policy — which was quite effective against some people.

At the time, I was conferring with J.P. and Randal about finding a way to get back on the air. We had actually entertained an idea of putting a legal shortwave transmitter in another country because we felt that a free, uncensored radio voice is necessary.

So far, nothing was going on with the ship. It was an interesting time. Freedom had been breaking out all over Europe. The Berlin Wall was now just a symbol. Oppressed European nations were calling out to be free, while here at home the government was using the "War on Drugs" to threaten all of our civil liberties.

I fear what may become of this nation. Too many people want to take away our rights. There are too many laws, too many regulations, and too much power for the centralized authorities.

I wanted to run a radio station dedicated to informing listeners of how to give something back to our country and how to preserve her. But I feared I might not get anywhere with the blockheads in Washington. I continued to strain my brain to come up with a safe, legal way to operate a station. Alas, it was difficult during those oppressive radio times.

Not a whole lot happened with the ship situation for the rest of the winter. During February, I traveled down to Texas to meet with John X and his partner. We spent six hours talking about offshore radio. Later, they told me about a deal where a country music group was thinking about putting up money to finance the ship broadcasting from here to Europe, selling cowboy items along the way. As unlikely as it sounded, I went along with it — but I doubted if anything would really come of it.

The FCC was processing my application for an international shortwave station very, very, very slowly. By mid-February, it made it as far as the hearings branch.

In April, the FCC told me that they were holding a hearing to determine my character. To me, as a person who holds the First Amendment Freedom of Expression very near and dear, the very concept that our government would hold a hearing about someone's character is preposterous — it's something that you would expect in a tyrannical, dictatorial-type regime. I know that a civil hearing is not quite the same as a criminal trial. But it is the same thing in your heart when you're up there sitting on the stand answering questions, knowing that your character is being judged based on what you say. It was a very upsetting thing.

(They were actually going to assess if I had the required character to be a broadcaster. Whether I do or don't I guess is for

God to determine. But it's just the thought of it. Most people don't even realize that the government does these things.)

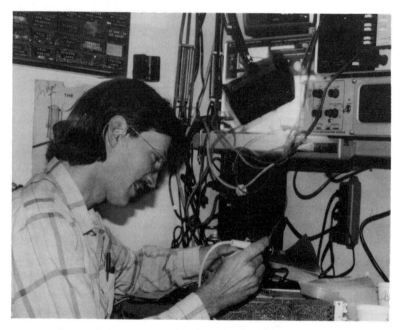

Allan H. Weiner at work at ABC Network Television, 1990.

My life at this time was working at ABC, spending time with Elayne, and fantasizing about something going on with the ship. The only thing I had to look forward to was a sort of Spanish Inquisitional hearing at the FCC regarding my application to build a shortwave station in Monticello, Maine.

On July 17, the big day arrived for the pre-hearing for the determination of my eligibility of character to obtain a shortwave license. It was also the day that I saw some real faces at the FCC. The surprise came when I saw a familiar name on an office door at the FCC's Enforcement Branch. Could it really be "The Fig" of WXMN in 1971?

It turned out to be true. I was somewhat surprised, but it's interesting to speculate on our lives and the small circles that one builds in life. He was an attorney in the Enforcement Division — the very branch that was hearing my case. In fact, he mentioned to me that he had to remove himself from having anything to do with that case because he knew me. Strange, but that's what happened. Aside from telling me about all our friends and the whereabouts of Larry Rand, he filled me in on some of the inner workings of the Commission. He said that they are a smaller organization than some people think.

I found out at the pre-hearing conference that a tough judge was being assigned to hear the case. Even at that time, I knew that they were going to do whatever they could to prevent me from getting a license. Kind of crazy, wasn't it? I went on a ship outside of the jurisdiction of the FCC so I could be on the air, and they slammed me off. I tried to get on the air legally, and they put me through all this hell. Life is just not fair, I guess.

Chapter Eleven
Radio New York International
On WWCR, WRNO And RFPI

In early August, 1990, Radio WWCR, a powerful shortwave station in Nashville, Tennessee, wrote me a letter suggesting that I lease time on their transmitter and bring RNI back on the air. I talked to their management the next day, and they seemed like a nice bunch of folks. So I called up the gang to see what they thought of the idea. Needless to say, Randal, J.P., and the crew were very excited. One moment, we were off the air with little hope of returning. Then, out of the blue, we were offered a way of going back on the air with 100,000 watts of power — legal and everything!

We had to come up with $15,000. A big bite of cash, but it had to be done. We'd have to order Comrex line extender and a bunch of other equipment. As soon as the contract was signed, we started promoting. It was hard to believe that RNI would be back on the air — on shortwave, no less! And with a lot of power, too. I hoped it would all work! If all went as planned, starting on September 16, WWCR would start leasing us air time on Sundays from 9 PM to 1 AM EST.

While I was dealing with all that, the FCC hit me with a list of documents that they wanted me to produce for the upcoming hearing on my shortwave license. They even asked me if I kept a journal. (Well, I guess I did!) I knew they'd just love to get hold of it. I hated to give them anything. Privacy is such an important right — a pity it's constantly being eroded away.

On September 9, John X and his partner bought the *M/V Sarah*. They gave me a $10,000 down payment, with the rest of the money contracted at $1,000 per month and no interest, for a total of $50,000. (Most of that $10,000 went to the ship yard and Frank to pay off bills.)

So the *M/V Sarah* was finally gone. We did our thing. I tried, but it was the end of another era. Would she ever broadcast again? That remained to be seen. There are always other ships, and there will always be another time.

On September 16, 1990, RNI WAS BACK ON THE AIR!

At 9 PM Eastern Daylight Time, we linked up with WWCR on 7520 kilocycles and returned to the air. We had a wild show, and no problems. The phone link-up worked great. We got so many calls that the phone lines were jammed. Randi Steel (Randal's air name) did a fantastic opening montage, and Johnny Lightning did a radio-comedy bit. It was a great show, a great time, and a great feeling to be back on the air.

We learned that RNI can be heard with a good signal in England, according to Chris Edwards of *Offshore Echo's* fame. Chris also informed me that Radio Caroline was not doing very well. They were fighting a battle with a land station on 558 AM.

Broadcast #2 over WWCR went well. Johnny Lightning did a riot of a show. He did some very good shows on RNI, even though he did tend to get a little bit risqué at times. Johnny always walks right up to that imaginary line where your listeners are either going to laugh their heads off or get ticked off because they don't appreciate what you're saying. But he always conducted himself well with RNI. So far, RNI's return had been a good one.

Johnny Lightning at his studio used for shortwave broadcasts of Radio New York International.

The United States government was trying to get us into another war that fall, this time in the Middle East. The habit of war must stop at some point in our evolutionary awareness. I hoped RNI could help. But meanwhile, I had a court date to attend and a shortwave license to fight for.

The FCC hearing started at 10 AM on October 4, 1990. Formalities took the first thirty minutes. Then I took the stand and the FCC attorney cross-examined me. He wasn't too bad. But the judge hammered away at me all throughout the proceedings. "Mr. Weiner," he kept saying, "Do you think you are a law unto yourself?" I was as apologetic and as understanding as I could be. I even said that I was sorry for all the trouble that I had caused the Commission. (It really hurt me to have to say that.) Once that part was over, I had to cross-examine the Field Agent from the Belfast Office in Maine. The proceedings ended at 4 PM.

*Elayne Star (Pirate Kitty) at Randi Steel's studio
used for RNI shortwave broadcasts.*

All in all, the hearing went as expected. Everyone said that I did the best I could under the circumstances. It was very reassuring to look out and see Elayne and Karl. Friendly faces in a room of stormy ones. My overall impression was that the FCC didn't like me — and they didn't want to give me a license. They would try to show deception, deceit, and disregard for Commission rules, and would insinuate that I was generally a bad person. I guess they don't have anything better to do.

I brought many letters of recommendation of my character to the hearing. One was from the editor of *Popular Communications* magazine, Tom Kneitel. He'd been in the business for a long time, and is so well-respected in the field that I thought his opinion would carry some weight. But instead, they took every word apart, bringing up the fact that he was indirectly responsible for editing the "Pirate's Den" column in *Popular Communications* magazine. I thought to myself, "So what? Tom's not breaking the law — he's just reporting on the activity of unlicensed radio broadcasters." The government people tried to discredit him, as well as just about everything I said.

I hired an attorney to do the proposed finding of fact, which we were supposed to submit by November 30. It required a lot of wording, and I felt I should have some legal counsel on it. In retrospect, I probably should have had an attorney with me at the hearing. Knowing how the FCC felt about me, the outcome would probably have been exactly the same. But perhaps I wouldn't have been treated so poorly by the judge.

I expected the FCC attorney to be nasty and caustic, but he was as gentlemanly as he was supposed to be. However, I was very surprised that the judge treated me as rudely he did. This was not a criminal court of law, so the rules were a little laxer. But when my attorney read the transcript, even he felt that the way the judge talked to me was not proper at all.

I don't know why the judge had such a negative attitude towards me. I never got angry or said anything nasty. I was very respectful towards everyone, conducted myself in a proper tone and manner, and was dressed as one should be when they go into a formal courtroom. If it had been a criminal procedure, I am sure that my defense attorney would have been able to get the case dismissed or thrown out, based on the prejudice of the judge — because it was pretty obvious that he had some prejudices. Now whether it was because I was a young person, my hair was a little long, or my name was Weiner, I don't know. But the Commission obviously selected a judge that was hard and nasty — and so was his decision.

We went to the first Monitoring Times convention in Nashville during the first week in October, and I gave my speech on pirate radio. There was a great crowd. I sold some RNI tapes and I spoke with a lot of people.

In mid-November, I spoke to Jim Latham, the General Manager of Radio For Peace International (RFPI) in Costa Rica. Building a transmitter site there was a possibility. RFPI was a unique station. It was built at the University of Peace in Costa Rica to promote the idea of global awareness and world peace. The university was built on land owned by the United Nations, making it a separate, "free" sovereign region.

According to Michael Couzens, the attorney I hired to help me with the shortwave license application, I was dead meat with the FCC as far as a shortwave license was concerned. Costa Rica had some advantages, mainly that they have cheap electricity and let you pick your own frequency. There were some drawbacks, though. It was such a long way from Maine and New York.

Eventually, the Costa Rican station failed to become a reality, mainly because of money and other difficulties. It turned out that the only way you could have a shortwave station in Costa Rica was to pay a Costa Rican to build the transmission site. But we did arrange to broadcast some of our RNI shows over RFPI.

I struck a deal with RFPI to get a couple of hours per week of air time, beginning on the first Saturday in January, 1991. When we first went on the air over RFPI, I sent them around $100 a month as a donation. In exchange, they let us have a few hours a week on 7.357 MHz and their various other frequencies. The programs we did for RFPI were recorded on cassette tapes, which we sent down for them to air. A live telephone connection to Costa Rica was too expensive.

But I still wanted to have a station of my own here in the U.S.A. My attorney suggested getting a trustee to be licensee of the Maine shortwave station until I prove myself as fit. When Michael posed that option to the Commission, they said it would just be a thin disguise, because ultimately, I would still be running the station. So we didn't get anywhere with that idea.

And I continued to work at ABC. That was frustrating in some ways. I was making $1,000 a week, but after the mortgage payments, utilities, transportation costs, and the horrendous taxes the government puts on a single person were taken out, it didn't amount to much. I only had an expendable income of about $20 or $30 a day to buy food, get around, and live on.

By the end of January, 1991, John X and his partner were four months behind on their payments for my ship. But I guess I kind of expected that to happen.

In mid-February, RNI was able to expand its programming. I struck a deal with Joe Costello of WRNO to buy a Monday through Friday spot from 9 PM to 10 PM on 7355 kilocycles. March 18 was a tentative start date.

I realized that $50 an hour was a lot of money, but I hoped it would help turn Radio New York International into a real seven-day-a-week alternative.

That winter, Elayne and I went through some rough times in our relationship and broke up. But even while I was in turmoil over breaking up with her, I was very happy that RNI was back on the air and doing what we could to stop the Gulf War. We were one of the few voices on the air at the time who were saying, "The war is wrong! It's a bloody mistake, and it's a waste of lives, time, and money."

Nearly everyone else in the media wanted to fight. They were beating their chests and saying things like, "Oh, let's have a war — Saddam Hussein is no good, so let's whip him." There were very few powerful voices of opposition, and I am proud to say that RNI was one of them. J.P. did an excellent job on the air of bringing up various points about the war, and so did Randi, Johnny Lightning, and everyone else.

At the time, we were debating marketing a whole line of RNI products like mugs, T-shirts and things like that — which we eventually did. We weren't trying to make a profit with RNI, we just wanted to make enough money to pay for air time and the expenses. But unfortunately, love, peace, and understanding — at least on the radio — does not sell very well.

RNI was now on WRNO. The audio was great, clean and crisp. It's amazing what a two-line Comrex can do. WWCR was no comparison. What was wrong with their audio?

On the downside, I was running out of money to fund RNI, and so was J.P. If income from donations and advertising didn't improve, the weekday service would have to go. It was sad that it had to cost so much. Five kilowatts, a dipole and a license were all we needed.

In early April, we got into a big argument with Joe Costello. When I was on the air one night, a caller accidentally used the F-word over the air. Of course, I cut him off. But Costello heard it and got very upset about it. He couldn't believe that we were running a telephone line directly to a 100-kilowatt transmitter without a delay. Of course, it's done all the time, but he insisted that if we remained on WRNO, we were going to have to buy delay lines, speech delays and all that stuff — or else he wouldn't let us take calls over the air. I was sorry that Costello felt that way, but the conflict over the speech delays ended our relationship with him, and we just decided to drop it.

So in the spring of 1991, I was single, alone, and broke again on account of radio — but basically content. My philosophy of life is this: You need food, shelter, warmth, clothing, and worship of God. If you have those basic things, you have it all.

Later that month, J.P. and I traveled to Carlisle, Pennsylvania, to attend an antique car show and swap meet. Along Route 78, we came upon Radio WMLK. We stopped by, met the people there, shot some video footage, and got a grand tour from the engineer. A unique station — built mostly by the people that run it. After seeing that installation, I was confident that I could build one for under $100,000.

In early June, a fellow from Radio Caroline called and asked if they could have some air time on RNI. I told him that we might be able to give them a half hour or so. Randi was against it, but I thought he would see it my way. At this time, Radio Caroline was off the air due to the new Marine Offenses Act. We eventually gave them some time. Whenever they sent a tape, I'd broadcast it — usually on my portion of the show.

There was a lot of turmoil at RNI, but we did have our glowing moments. The programs we aired every week were a lot of fun, and they were always interesting. There were moments when we all got along in harmony. But almost always, there was an undercurrent of problems, mostly between J.P. and Randi.

By the third week in June, RNI had officially broken up. It was such a pity that it came to that. For some reason, J.P. and Randi could not compromise or have any trust in each other at all. Perhaps the FCC did not really have to close us down at all. We could do that by ourselves.

Reflecting on it almost five years later, I can really understand Randi's frustrations in dealing with J.P. Even to this day, I have never been happy with the way things turned out. When Randi left, a lot of the spark of that organization went with him.

Randi blamed J.P. for just about everything, including the loss of all my licenses, and the fact that the FCC wouldn't license me for a shortwave station. I guess there was some merit to that. However, I have only myself to blame, as I made all of the final decisions. No one really twisted my arm.

RNI was soon restructured. I picked up the Comrex equipment after turning over $2,300 in cash to Randi. I also got some RNI mail from him.

I never meant for any of this to happen. RNI was supposed to be a free radio station, made up of people doing their thing, held together by the common love of the media. It was supposed to be fun. I really thought that people could get along. But I guess it was not to be.

J.P. and I decided to run RNI by ourselves. Maybe it wouldn't be perfect, but at least we'd have some fun. I'd had a great time being on the air. True — I messed up here and there, but I knew that the audience still enjoyed the show. Our purpose was still the same: Spread the words of love, peace and understanding. In a time of violent aggression, everything helped.

On July 19, 1991, the judge in the FCC's case against me rendered his decision of "no," as expected. He denied me a license based on the grounds that I did not have the requisite character to be a licensee of the Federal Government. So now we would appeal, most likely to the full Commission.

Meanwhile, the *M/V Sarah* was still rusting nicely in Peter's boat yard. And John X's partner said that they would start paying me in August.

Our programs on WWCR were calm, peaceful and fun. Even though I missed everyone from the old gang, I must admit that it was nice to have RNI peaceful for a change. It might not have been the best programming on the air, without all of the diverse voices we had before, but there were no more arguments or fighting. J.P. and I more or less got along. And after a few weeks, Johnny Lightning and some of the others decided to return to RNI. I was very happy to have them back on the air again.

Meanwhile, another problem was brewing. Because of the damn FCC and their recently published decision about me, I had to attend a stupid meeting at ABC on August 1st. They had this inane worry about conflict of interest — as if anything RNI did on Sunday nights could cause conflict to ABC.

(When the FCC published their decision regarding my case, it was put into the public register. Then some of the legal eagles at ABC found out about it, and memos started flying around my department that said they had a rogue working for them who was just denied a shortwave license by the FCC. A few people at ABC got nervous about it and started calling us into their offices.)

The nuts and bolts of the situation at ABC was this: J.P. is an excellent engineer, and he was working up at the test equipment shop doing one of the best jobs they ever had. And I also did a good job. I made repairs on equipment, installed things, and made things happen.

Then that stupid decision regarding my shortwave license came along, and the bureaucracy at ABC went berserk over it. Maybe it's because I'm Jewish — I don't know what the hell it was. But it was really annoying. At that time, ABC was paying people who found a corner to sleep in every day instead of doing any work. They'd been getting by with it for years and years — and those people were still there. But I dared to apply to the United States government to build a shortwave broadcasting station and put some alternative radio on the air, and the brass at ABC started giving me flak about it.

We had a meeting at the union headquarters, and I was furious about the whole thing. At one point, I said to J.P., "I don't care if I stay here or not. If you want to stay, I'll admit to everything and lobby for your situation." But before long, J.P. and I both decided that we just didn't want to work for an organization that treats people that way. So we resigned. Looking back on it now, one of the best decisions I ever made in my life was to get the hell out of that pit.

We signed the termination papers on August 7, and left ABC for the last time. We then went back to Yonkers, jumped in my boat, and went sailing for the rest of the afternoon. It was one of those good days you make out of a bad day. We were looking ahead and talking about other businesses that we could get into — and exploring the possibility of buying a small-town radio station.

We met with ABC again on August 15, and settled our benefits. J.P. and I felt fine. Now it was on to buying a radio station, doing RNI Surplus Electronics, selling shortwave car converters — or whatever. I was very happy, because my time was once again my own. To celebrate, I went to Maine.

August 17, 1991. My life up to that point: first I went to Maine, then I met Sarah, then I played radio and TV at Ricker College, then I built Uncle Fred's natural food store, then I built a print shop, then I bought WELF, then I built WOZI, then I experimented with KPRC, AM and shortwave and KPF-941, then BOOM! Ships at sea, destroying my back for a year, offshore Long Island, NY, then BOOM! Then working at ABC, the return of offshore Radio New York International, tons of money spent, and then BOOM! More ABC, Sarah leaving, Elayne coming in and out of my life, settled into the dull existence of an 8-hour day at ABC, then BOOM! And so it goes.

Chapter Twelve
J.P. And I Buy A Small Town
Radio Station

In the late summer of 1991, J.P. and I started looking in earnest for a small town radio station that we could afford to buy. We checked out quite a few stations throughout the New England area, but most of them were overpriced or had so many problems that we knew it would be impossible to make a profit with them. So we just kept looking.

We looked at WHVW in Hyde Park, New York, in August. They wanted a lot of money for it. It was a losing AM station, but J.P. liked it a lot — mostly because he liked the area. It would be a big challenge. But if nothing better turned up, it might be a possibility.

Later that month, I took a vacation trip to Europe with my father. We had a good time touring Finland, Sweden, and Russia. Warm weather greeted our return to the USA, along with news that J.P. did not watch my sailboat and it nearly sank. The engine got swamped and oil was everywhere. I wanted to go to Maine, but instead I had to stay and clean up the boat. After many hours of cleaning, pumping, and repairing the *Bob*, it was back in running order. The old iron gas engine fired up on the first turn.

J.P. called Don, the owner of WHVW, and offered him $280,000 for the station. The owners wanted $350,000 for it. Too much! So J.P. kept trying to negotiate with them.

In late September, I picked up James Latham, the General Manager of Radio For Peace International at the airport. We talked about radio all day. RFPI, born in the freedom of neutral land in

Costa Rica, is quite a station. Unique in its own right. It was a most interesting visit.

All of our problems and frustrations at RNI were due to the fact that we didn't own a transmitter site. On this, we could all agree. But what to do about it — another country, another ship, make our own country? Maybe I could start a sovereign state somewhere. I knew that the United States owns the world, but perhaps not every single piece of it.

James Latham was great to talk to. We shared a lot of the same notions. Also, we had similar experiences in radio.

Explaining RNI and all of its history was difficult. After 20 years of dealing with this crap, you would think we'd have half a billion watts on the air. But no, just more vindictive aggressive onslaughts from the FCC. Why are they forcing me to broadcast from outside of my own country?

Not long afterwards, Tom Kneitel of *Popular Communications* told me about a piece of land where the sovereignty was undetermined — Machias Seal Island, off the coast of Maine. The problem was finding someone neutral to ask the right questions.

The sovereignty of that island has been in dispute for years and years. Canada has a lighthouse station on it, and the United States government also feels it should belong to them. I thought about putting a station there for a while, but I came to the conclusion that it would be very expensive to build — and chances are, both governments would come out and destroy it.

Also, a couple of islands northeast of Nova Scotia are officially under French protection. Perhaps shortwave transmitter sites could be had there. Crazy idea, but those were times for way-out thoughts anyway. Freedom of the air must be preserved. Besides, I'd always wanted my own country.

I also wanted a ship, a tall ship, a vessel with masts that can fill with air and move with the wind. One that is 80 feet long, easy to handle, and equipped with at least 10,000 watts of shortwave transmitting equipment, ready to broadcast.

In early October, like an idiot, I got involved in a love triangle. I didn't mean to, but it happened anyway.

J.P.'s girlfriend, Julie, called me up and invited me out to dinner. So I said fine. A little later, J.P. came over to the house to discuss some RNI matters and I asked him, "Hey, what's going on between you and your girlfriend? She called me up and said that she really wanted to see me — and I have a feeling that there's more to it than just going out to dinner."

He said, "We broke up about a week or so ago. There wasn't really any relationship there in the first place. So whatever happens, go for it. Have a good time."

I asked him, "Are you sure?" Because my philosophy of life is as follows: You don't fool around with other people's girlfriends or boyfriends, and you don't fool around with other people's wives or husbands. I'm very adamant about it — that's how it has to be if you want to live an honorable life. So I was very cautious about the whole situation. And I asked J.P. two or three more times to make sure he really meant what he said.

Quite honestly, I was very lonely and ready for any female company. So if J.P. had no problems with it, Julie's invitation was certainly welcomed.

In the following days, Julie and I spent quite a bit of time with each other. I was starting to get infatuated with her because she's a nice person, we got along well, and we thought we really liked each other.

Then J.P. came to me out of the blue and said, "Wait a minute — Julie is my girlfriend. Why are you doing this to me?"

If there was any point in my life where I really got wrecked (or "zinged," as I called it), that was one of them. Sarah leaving me for a woman was bad enough. But then for my best friend, of all people, to do this to me was unbelievable. One minute he told me that he wasn't going out with Julie any more — then after I really got to like her, he demanded that I give her back to him.

J.P. told me that he only said it was okay for us to date so that he could test Julie. I was stunned and emotionally shocked. I looked at him and said, "You used my heart and my emotional situation to test your supposed girlfriend?"

J.P. knew very well that I'm a sensitive person when it comes to relationships. I do not take them lightly. And when I get involved with someone, I spend a lot of time with them and nurture the relationship. I do the compromises and I allow the relationship to grow as far as it can grow.

Nothing like that had ever happened to me before, and I certainly didn't expect it from my best friend. Here we were getting ready to go into a business deal to buy a radio station, and he did that to me. It really hurt a lot. And it took me months and months to heal from it.

J.P. was angry at the way his experiment turned out. We were still friends. But after all that had happened, I wondered if things could ever be the same between us.

On top of all that, the WHVW gang was being difficult. They really didn't seem to know what kind of deal they wanted for the station.

In mid-October I had a thought: Maybe I should try to get a shortwave license in Monticello assigned to J.P. This had been discussed before; however, it seemed to be a viable alternative. If it could be done, the shortwave facility would be a business for both of us. We could run Radio New York International on a daily basis and lease the rest of the time out. WWCR and WRNO seemed to have little problem finding people to lease time on their transmitters. So why not us? When I returned to New York, J.P. and I would fill out an application and have Michael Couzens submit it to the Commission. As long as my name wasn't on it, there was a chance it could go through.

Then J.P. called me a couple of days later to say he could not stand the thought of me having a relationship with Julie — even though there was a fat chance of that after all that happened. If we did, he said he would have to give up RNI, WHVW and any other radio deals we planned — as well as our friendship. All for a woman that he left and told me to go for. You go figure. But deep down inside, I knew that J.P. and I would somehow remain friends. I realized that he was in some pain also. I just wish that he had not let this confusing cloud of thought undermine our two decades of friendship.

This really affected the way that RNI was going. In fact, we almost lost RNI on account of the whole mess. It caused a serious rift between J.P. and me for a while. And it changed our relationship forever. Even to this day, I can't really look at him the same.

In his lifetime, J.P. has had dozens of girlfriends that lasted anywhere from an hour to a year. But me, I have relationships that last for years and years. So I honestly don't think he realized how much that incident hurt me.

In late October, I spoke with Michael about getting J.P. licensed for the shortwave station. After some hesitation, he agreed. He would talk to the FCC and see what they thought. I hated to go this route, but what else was there to do?

But, the idea of getting a shortwave license in J.P.'s name went nowhere. "It would be like trading Hitler for Attila the Hun," the FCC said. So that was the end of that idea.

In late November, RNI went up on the satellite via the Becker Satellite Network. We were on Spacenet 3 transponder 21.

J.P. and I visited WARE 1250 AM in Ware, Massachusetts. A strange single-station market. It should have cost around $100,000, but they wanted $400,000. Too much!

In early December, I spoke to one of WHVW's owners again. They finally agreed to sell the station to us for $100,000 down and an installment sale on the $250,000 balance on the property.

J.P. and I saw Pop on Friday. He seemed very happy about the whole thing. Later that day, we signed a purchase agreement and faxed it to the owner. We went to Hyde Park the following week to sign the papers and give them $20,000 in earnest money. My father and I made a personal loan to J.P., which he was obliged to pay back to us. Michael Couzens saw no problem with J.P. applying for the WHVW license.

J.P. and I signed the agreement to buy WHVW on December 13. Finally — after four months of waffling around. So pending FCC approval, we would soon be stuck with an AM nightmare. Actually, it looked like fun, but it would be a challenge to turn it into a profitable contribution to the mid-Hudson community.

The October surprise continued to put a strain on relations between J.P., Julie, and me. So, in mid-December, we had a meeting to discuss the situation. I realized that what had been going on here was very bad. I said to Julie and J.P., "Why don't you two get back together, go about your lives, and work it out? I don't want any more phone calls — I don't want anything to do with it. I have radio stations to build and all kinds of other projects to work on. Just leave me out of it."

After that, Elayne and I immediately got back together. On Christmas, we secluded ourselves at the Bear Mountain Inn in Bear Mountain, New York, and spent a couple of days just talking, soothing each other, and healing. I really fell in love with Elayne all over again. And from that moment on, we have been together.

That was the resolution of the issue that almost destroyed me, RNI, the friendship between myself and J.P., as well as my relationship with Elayne.

One important thing I learned from the incident is that there really is evil in the world. Whether it's force, Beelzebub, Satan or whatever you want to call it, evil really does exist. And if you exclude God or righteousness from your life and accept a path of evil, as many people do, it really consumes you like a fire. It can destroy you and everyone around you — especially those who love you very much.

By early January, 1992, I was up in Maine trying to help Dr. Rish get WREM back on track. How to run a broken-down AM radio station in the middle of nowhere (Monticello) and make a profit? So far, a country music format was about the best idea I could come up with. I thought we would go to tape automation. The only question was — should I use reel-to-reel, cassette, or VHS hi-fi?

I saw saving WREM as a challenge. Who else would attempt to salvage the unsalvageable? It would be good practice for when J.P. and I began to run WHVW — hopefully in the spring.

On January 22, I installed a studio in my friend Seth's jewelry store in Houlton. I did pretty well for six hours of work. I put in a nice 8-channel console, a couple of tape recorders, a CD player, cart machines, cassette machines, and linked it by telephone circuit to the transmitter here in Monticello. When it was finished, Seth could control WREM's programming from his store.

Seth's store seemed to be frequented by a wide assortment of nuts, misfits, and weirdos. I should talk. But, I knew this could be a fun radio station. That winter I kept busy building, repairing, and doing radio. And I loved it. It was the best therapy for me.

In late January, my attorney called and told me that the FCC Appeals Board denied my application for a shortwave license. They claimed that I can't decide if I'm Peter Pan or Captain Hook. All I wanted to do was legally build a shortwave station at my farm in Monticello, Maine — and that's what my government tells me.

WREM was coming along nicely, though. The audio on WREM sounded good. All this over a telephone circuit. The remote control system was working out well, also. The studio-in-the-store concept was really interesting. It certainly provided much better exposure. And I hoped it would sell some advertising.

RNI was still moving along slowly. We had to give up the satellite link and go back to the phone hookup with WWCR. Financially, it was a drain. There was slow improvement, though. If I could only turn my efforts into monetary value — I'd have to keep on trying.

On Friday the 13th of March, I had to take J.P. to the hospital because he was having severe abdominal pains. After waiting two hours in the emergency room, he finally got some medical attention. First they thought appendicitis, maybe a urinary tract infection. Then they decided that he had kidney stones. After a total of eight hours in the emergency room, they finally admitted him, drugged him and placed him on an IV. In a few days he felt better, and they decided to release him.

On March 10, the FCC approved the transfer of WHVW's license to J.P. The amazing thing is that they did it in less than two months. Astonishing!

So here we went again! Another radio station to try and make a go of. This one, however, was established, and it did have a base clientele. I'd have to trust J.P. as far as the station was concerned. Pop and I were putting up all the money as a loan to him. I realized that J.P. and I went back over 20 years and we were brothers in this, but you know what happens when money and things of value are involved — not to mention relationships and women.

The October incident still smarted a bit. But hopefully, J.P. would be honorable, as he said he would. Dad was happy about the FCC decision, and he was also very supportive. Coming up with the money to buy this station was no small order.

The deadline passed on John X's partners' payment on the *M/V Sarah*. I'd have to hire an attorney in Boston to repossess it — but at the moment, I didn't have the money. When we got WHVW on track, maybe my finances would smooth out a bit. All I had to do is stop buying so much collectable equipment. (I buy old radios, TVs, radar equipment, and other items of technology. Over the years, I've filled up about a dozen buildings with that kind of stuff.)

WHVW was about to go bankrupt. We had half a dozen liens against the station, and the whole staff was threatening to quit because the previous owners hadn't been paying them.

On top of it all, Elayne said that our relationship would be over as soon as I started working with all the creatures at WHVW. Elayne really felt that if I got involved in a business with J.P., it would be a dismal failure. And she was upset with me for not listening to her.

In early April, RNI's satellite service with Scott Becker was back on. Johnny Lightning called to say that he had to come up with a $1,750 application fee for the shortwave license.

Pop, J.P., and I went to a pre-closing meeting at WHVW. It looked like we could take over the station on Thursday, April 16. We had a staff in place — plus some ad accounts were currently running.

I knew that getting the station on track would be a difficult uphill climb. This time, however, with a partner. Again, I had no money and a broken radio station. There was a good chance for this one to make it though. So let's give it a try!

On April 16, we bought WHVW. I handed over $100,000 in a loan to J.P. Then J.P. signed a bunch of papers and BINGO! — another station to deal with. No money as usual, but we planned to build up cash flow with increased sales.

Almost as soon as we bought WHVW, J.P. and I got into some disagreements. J.P. wanted WHVW to have a mixed-bag rock'n'roll format, but I wanted it to go country. I felt that country music was the only salvation for a stand-alone AM station in a well-represented market where there were seven or eight other stations.

But J.P. thought that country music was trash and he didn't want to have anything to do with it. So, against my better judgment, I decided to let him try his idea for a while and see what happened.

WHVW was a dump. We had hornets in the office and snakes in the basement. Even the toilets needed to be worked on. On top of that, we had problems with the automation equipment and couldn't find the right satellite service.

The staff was mad about not getting their back pay. By the end of the first week, it would most likely be J.P., myself, and Ken, our sales manager. There was little money and few advertising accounts. What a station!

Everything that possibly could, went wrong at WHVW. The staff was still unsettled and the new format was a bomb. Listening to J.P. was my biggest mistake. So I forced the country-format issue and finally won.

I sat down with our sales manager and said, "This station should go country. It's the only format that has even a remote chance of gaining listeners and making advertising bucks." He was into country music, and agreed with me. But J.P. was fighting and scrapping every inch of the way. Eventually, J.P. came to love country music. But at the time, country music was like dirt to him and he wanted nothing to do with it.

Aside from this, there were still thousands of problems — not to mention cash flow and little money to make payroll. It was just like WOZI all over again — this time with an impractical partner. But the mortgage papers had been signed, and Pop and I were financially committed. So I had to give it a shot.

The entire air staff quit on April 22. It was sort of a surprise, but not completely unexpected. But J.P. was finally beginning to think like me. He agreed to cut the number of employees down from 12 to about 4 or 5 and go with the country format.

Soon after we took WHVW over, we found out that it wasn't making the kind of money that was stated in the original prospectus. The cash flow and total income of the station was listed as around $17,000 a month, but it turned out to be more like $7,000. And that really made things bad. It eventually led to legal action, which got us released from a lot of financial obligations. But in those first formative weeks, it was very hard for J.P. and me to run the station because the cash flow was so low.

Driving and radio became my life. Yonkers is 75-80 miles away from WHVW, and I had to commute every day. We found a morning man who wanted to do the show for nothing — my good friend, Tom Barna. My sister, Barbara, was also helping out with the station. I had to move the *Bob* up to Hyde Park as soon as possible, and I needed a place to stay.

It was nice to be working again, building another radio station. Speaking of radio stations, I needed to go to Maine as soon as possible to fix up WREM. Seth said that the station was doing well — however, there were lots of technical problems. But I didn't dare leave until WHVW was on the satellite service playing country music. It was a damn shame that we had to waste so much time trying that other format.

I repaired cart machines, remote units, and production equipment, installed lamps, fixed pipes — there was so much to do at this station! I was living on the *Bob* for a while until I found a place to stay. I also needed to find a place to park it. Hyde Park had a marina, but it was expensive.

WHVW had live programming in the mornings and in the afternoons — when J.P. did his show. We only used the satellite service for middays, in the evening from 6 PM to 6 AM, and on the weekends at certain times. The weekends were filled with all kinds of specialty programs. We had an Italian program, a German show, an Irish show, a gardening show, a pet show, and some religious programs.

WHVW crawled along. Somehow, more and more people were working there. The payroll had grown, but J.P. and I still received no salary. It was the same with my sister, who worked at the station about three days a week. Country music was still a hit, and ad sales seemed to be creeping up.

Even though things were going a little better at WHVW, a lot of interpersonal problems were starting to develop. J.P. and I got along pretty well, but when I tried to do my job as general manager and advised him on programming and how things should go, he sometimes gave me trouble. The main problem that I was having with J.P. was that WHVW wasn't a pirate station. It was a commercial enterprise with a payroll and a big mortgage payment that had to be made every month. And to do that, we had to make decisions in a businesslike way.

Another problem was that our sales manager didn't really want to sell advertising — he mainly wanted to be on the air and do promotions. But he didn't tell us that, and I wasted $12,000 or $13,000 paying this guy to do stuff that really didn't need to be done. Finally one day, I took him into my office and dismissed him.

We went through one sales person after the other. Then we hired a gal named Cathy, and she turned out to be one of the best salespeople we ever had. The only problem was that J.P. decided to seduce her before he even hired her, which made things difficult.

The rule is: don't get involved with your employees or it will compromise your position. I told him not to get involved with any of the people we hired — especially not in the formative stages of WHVW's operation. But he didn't listen.

After he started going with her, Cathy took it personally whenever J.P. asked her to do certain things at the station. And I was always stuck in the middle of their disagreements. That was part of the reason that I eventually decided to sell out my interest in the station.

Meanwhile, I planned to take a trip to Maine. I had WREM and some farm matters to attend to. Seth and WREM seemed to be doing well. His laid-back radio style was different, as were his ad-lib advertisements. Considering his low overhead, it seemed to work.

J.P. decided to rent a cottage north of Hyde Park in Stattsburg, New York. After over 15 years of living on Saratoga Ave., in Yonkers, he escaped to the country. It was good to have him close by the station.

Elayne broke up with me again in mid-July. I was broke, WHVW was a drain, and RNI had no money either. What a mess! One day, I saw a bumper sticker that read, "God permits U-turns." Was it time to turn around and start again?

In other news, I heard that the FCC was now fining radio pirates up to $20,000. That sure is a heavy price to pay for such a victimless crime. According to Dick Smith of the FCC, I am in retirement from pirate radio. Gee, am I eligible for retirement benefits?

Elayne and I made up during the first week in August. Despite our differences, we do love each other very much. And we still wanted to be together.

Meanwhile, things were still going downhill financially with RNI. On September 6, 1992, we did our first fund raising marathon. I received a bill from WWCR for $4,500 the day before. With no money, we had to try something. We raised over $2,000 in pledges that one night. I was amazed. Even if we only got half of it, that would have been a big help. Johnny Lightning was going to match half of what we raised — that gave us $500 extra on the top. There WERE real listeners out there. And RNI would survive!

That fall, WREM's transmitter crapped out because of bad tubes, and I had to locate some new 828s and 814s. Michael Couzens filed Johnny Lightning's shortwave application. And I stayed on the farm, stayed in touch with the station by phone, and worked at repairing WREM.

Through donations RNI raised about $3,000. Still, we had a lot of bills to pay. RNI needed to stay on the air and expand into a network of alternative programming. With Rush Limbaugh and other extreme right-wing types on so many stations, there needed to be some balance to this crazy New World Order.

On October 12, I had a close call. I forgot to turn off the plate voltage before I stepped inside of WREM's transmitter. I opened up the modulator cabinet, walked inside of it, and suddenly realized that I hadn't shut the power off. I just froze, then I backed out very slowly, being careful not to touch anything. Considering that there were 9,000 volts surging around in there, it could have made a real crisp hotdog out of this Weiner! But after replacing many capacitors, resistors and tubes, WREM's audio was the best it had ever been.

By mid-November, WHVW was going down the tubes. Austerity measures would have to be taken real soon if sales didn't improve. Cash was low and J.P. was about at the bottom of his reserves — plus his car blew a ring or something. It was NOT supposed to be like this. Later that month, I was finally able to repossess the *M/V Sarah*. I immediately put her up for sale again.

Together with the usual transmitter problems at WREM, Seth was about to toss in the towel. Few accounts, little money, and a backlog of receivables made it tough for him. Let's face it — an AM daytimer in northern Maine had little in its favor.

Christmas Day rolled around, and Elayne and I celebrated the birth of the Prince of Peace and had a nice day together.

A few days later, Seth told me that he wanted to give up on WREM. He was burned out. Another failure? I didn't know what to do. I was tired of the whole affair. No one seemed to care about local radio anymore.

I wrote to Anita McCormick, a dedicated RNI listener, and asked if she had any interest in helping me write my life story. I know — it was an egocentric thought. It might make good reading, however. Certainly there are enough weirdos, misfits, and malcontents in my life to make it interesting. Anita agreed to help with my story. Hopefully, someone would publish my account of my broadcast career.

1992 sure whizzed by. Not too happy of a year, I must say. As for New Year's resolutions, I didn't make any. But as always, I would try to be good! As I spiraled down that poorly lit tunnel, I looked for light along the way.

My friend, Tom Barna, decided to move into the mobile home of WREM. He said that he would build an automation system and help Seth run the radio station.

John X and his partner had been giving me trouble about *M/V Sarah*. After I repossessed it, they went crazy. I called them up and said if they paid their bill off, they could still have it. But only time would tell what happened next with them.

That winter, Pop invited me on a 10-day vacation in the Caribbean.

I had a fantastic trip and arrived back in New York on January 24. It was amazing how much junk mail accumulated while I was gone. Plus, there were 20 minutes of phone messages on my answering machine. J.P. said that WHVW was doing okay. He's making payroll, but not paying bills. With five salespeople and a great country-radio format, one would expect the station would at least make expenses.

Tom arrived at WREM to find a frozen studio and a cracked toilet. Apparently, Seth let the fuel run out. So I headed back to the farm to help Tom fix things and get the station back on the air.

In spite of all my efforts to save it, RNI was dying. It was too costly, and there wasn't enough support.

On February 19th, I asked WWCR to reduce our rate by 50%, and they agreed. So RNI was saved for the time being. We were the station that would not die.

(We had been paying $60 an hour, and I asked George to reduce it to $30. He said that he would, but he also told me that if someone else contacted him that was willing to pay the full price, he'd have to give us first option to pay the original rate or give it up.)

WHVW broadcast President Clinton's address at the Havalin Middle School when he came to Hyde Park. I got to see him in the auditorium. Secret Service men were running around everywhere, and there was an army of press people and cheering crowds.

In late February, some people in the movie industry approached me about renting them some of my early television equipment to use in a picture about the quiz programs of the 1950s.

That project turned into the *Quiz Show* movie with Robert Redford. The moviemakers decided they wanted to rent a whole truckload of stuff — mainly old television equipment from the late '40s and early '50s. I had tons of it because I'd been collecting the stuff for nearly 30 years. When the stations were throwing out to modernize, I went down with a pickup truck and got it for nothing.

I got an offer on the *M/V Sarah,* but no deposit. I'd been thinking about buying an old lightship from Miller Yacht Sales and converting it to an offshore shortwave broadcasting station.

Things were going a bit better at WHVW. WREM, however, was still on the verge of collapse. But for some unexplained reason, I felt optimistic that things were going to turn around soon. Pop had been wonderful in helping me out a bit with my dismal finances. But independence was, of course, my goal.

Chapter Thirteen
The Fury 5 Project

In early March 1993, Scott Becker contacted me and asked if I was interested in building another offshore radio station. He was very hot on the idea and, naturally, it sounded great to me too — especially since Radio New York International was in imminent danger of losing our Sunday-night time slot on WWCR. But I told Scott if we did this, everything had to be 100% legal and the ship would have to be licensed to broadcast from the territorial waters of some friendly nation. Scott agreed, and he immediately started looking for someone to finance the project.

On March 8 we spoke with Brother Stair of the Overcomer Ministries[1] about financial backing. About $150,000 should do it — not even the price of a 3-bedroom house. He said he would pray on it. So I would try to get a ship and hope for the best.

The next day, two bad things happened to RNI. First, WWCR canceled our program. Second, the FCC decided to give Johnny

---

[1] Brother Stair was the head of the Overcomer Ministries, a group of people who have come together to worship God and do his will. They had a 70-80 acre farmstead near Walterboro, South Carolina. Brother Stair believed that he was the last-days prophet of God. He said that God spoke to him regularly and told him that the end of the world was coming very soon, perhaps before the end of the century.

Brother Stair has a shortwave ministry and leased time on a handful of international broadcast stations — including WWCR and WRNO. We got to know each other through Scott Becker, the owner of Becker Satellite Systems.

Lightning a lot of trouble about getting his shortwave license. The offshore idea was sounding better all the time.

Three days later, I drove to Boston to see the *Fury 5*, a 140-foot dragger ship. She was wide, deep, and fully operational — almost turn-key. We heard that the owners would take about $45,000 for her. She had 10,000 gallons of fuel on board and was already in Boston — an almost perfect boat to use as a radio ship. It was the toughest-looking ship I'd seen. I arranged to speak with the owner and try to work out a deal.

The Radio Caroline people said that they'd try to provide some labor and crew people if I managed to get another offshore station together. All we needed to do was get a few more religious organizations to help with the cost of the construction. Scott was really excited about the whole thing, and so was I. So off we went again!

On March 18, I got a commitment from Brother Stair. He said that God told him to go ahead with the project, and he pledged $20,000 to buy the *Fury 5*.

We agreed to do the project in the following way: Brother Stair would finance the operation of the ship. We would have four shortwave transmitters on board. One would be totally dedicated to the Overcomer Ministries, and the other three would be leased out. And of course, RNI would use one of the transmitters to relay our programs.

We did have some trouble securing the ship at first. The fellows who owned it wanted too much money for it. I think we eventually offered them around $40,000, but they wanted a lot more. So I left the meeting and drove back to Monticello. When I got to the farm, I thought of a different approach they might go for.

I called up the owner and said, "Listen, we really don't have much money. So why don't we do this — we'll give you $20,000 to help pay off your immediate bills and then give you a percentage of the operating profits once we get it on the air?"

They agreed, but they wanted 50% of the profits. That sounded a little high. But since we were low on cash, we decided to accept it.

On April Fool's Day I secured the deal for the *Fury 5* — seven years ago to the day that I bought the *M/V Sarah*. Funny how these things occur. Now began more ship madness. Transmitters, antennas, generators, fuel, welding, iron, water, etc. It was very important to keep this project 100% legal.

I found out that WWCR burned to the ground on April 3. No one knew what caused the blaze.

Scott Becker and Brother Stair were really hot to trot on this project. My life would be getting very busy real soon. Most likely, there would be little time for hobbies and the like. But when I build, I am very happy. I told Elayne about the offshore station, and she was concerned that I might sail away forever. But she would participate, none the less.

Johnny Lightning and I visited the *M/V Sarah*. He cleaned, while I got the power on and the toilets to flush. Then I began to salvage the *M/V Sarah*. I'd decided to scrap the old radio ship. I would remove her anchor and chain, electrical items, all the radio stuff, and the 60-kilowatt generator. Johnny Lightning was a great help that week on the ship. Having someone along to help was a big asset.

I landed a deal for four shortwave transmitters from Radio Research Instrument Company, one 10 kilowatt set and two 40-kilowatt sets from the government. They seemed to be nice, compact units. Brother Stair came through with the money we needed to buy them.

The transmitters arrived in Boston about three weeks later. Peter Bang, Tom Barna, and I loaded them on the *Fury 5*. It was a hair-raising spectacle, to say the least. I also got the main engine to turn. And Scott Becker was out there somewhere raising money and doing some other stuff with Brother Stair.

WREM was off the air again. Seth threw in the towel because people weren't paying their bills for advertising — mainly on account of the bad economy.

In other news, WHVW's sales were up, but cash flow was still dismal. And my house at 14 Prospect Drive was now up for sale.

Since I was no longer working at ABC, there was no need to keep it.

Barbara squealed to Dad about the offshore project, and I was very annoyed at her. Dad did not like the idea of offshore radio, mainly because he was worried that I'd get shot or torpedoed by some government.

Meanwhile, I went down to the ship and started getting the transmitting equipment in order. Tom said he would come down and help. He built a nice studio for the station. And a guy named Herb said he would work out a deal to take care of the plumbing and mechanical stuff.

By the end of June, my life was as follows: Tuesday afternoon through Friday night, I worked at the ship. On Saturday and Sunday, I visited Elayne. Then the cycle repeated itself. Anyway, the ship was coming along fine. The transmitters were there waiting to be installed, and I was still working on the antenna system.[2]

MGM Pictures bought the *M/V Sarah* so they could blow her up in a movie they were making about a wild psychopathic bomber in the Boston area (the movie turned out to be *Blown Away*). I was sorry that she would be destroyed, but at least I'd get some money out of it to pay off the bills.

By mid-July, the transmitters — four shortwave, one medium wave (AM), and one FM — had been welded in and installed.

On a ship, transmitters have to be very well secured. You just can't stick them in a corner like you do on land. You have to weld them to the superstructure of the ship — because in rough weather, the sea tosses a ship around like a cork. And if the equipment isn't securely welded to the ship, it will break lose and start flying around the area. Even equipment that weighs thousands of pounds can slide around when the water gets choppy.

---

[2] The antennas I put on the *Fury* were untuned sloping cage antennas. I tapped them about ¼ of a wavelength from the stern. I had two of them— one strung to the port side and one strung to the starboard side of the ship from the forward mast. From a rigging standpoint, it worked out very well. We used the towers from the *M/V Sarah* to support them.

Brother Stair visited and did a live broadcast over WWCR via telephone. He seemed to be impressed with all the work we'd done on the project. But he didn't want us to work on Saturday, the Sabbath — so we didn't. Brother Stair was really quite a fellow. He was very sincere, knowledgeable, and fascinating to talk with.

Scott Becker went down to Nevis and St. Kitts to see if they would give us permission to broadcast from their waters. So far, all was favorable. I located a generator that would put out about 225 kilowatts to run the transmitters.

It was 90 degrees in early July. On the deck of the *Fury*, it got up to 100 degrees. I'd been busy working on the antenna system, and it was coming along fine. We were using the towers from the *M/V Sarah* to support it.

Right in the middle of all this construction, Dad came to me and said, "I want to take a 10-day cruise in the Mediterranean, and I'd like for you to come with me." It couldn't have come at a worse time, but how could I refuse? So I agreed to go to Europe with Pop on a cruise ship.

On July 24, my father and I left by plane for the *Pacific Princess*, which was docked in Barcelona, Spain. Scott left for Nevis/St. Kitts. He hoped to close the deal so we could bring the ship down there as soon as it was finished.

Pop and I visited Rome. The Vatican was big and beautiful. We did a lot of sightseeing — the cradle of civilization was awesome. We also went to Cannes, France, where we visited a few museums and shops.

At Gibraltar, we saw the rock, drove around the rock and then through the rock. In fact, the whole city was located in and around the rock. The weather was absolutely splendid. Pop and I had a fantastic time, and food was everywhere. When we returned from the cruise, I went to the ship, did some work in the transmitter hold, and then returned to New York.

Work on the ship was all but at a standstill. Engine room problems were holding us back, and only a trickle of funds were coming in from Brother Stair.

Like many radio ministers, Brother Stair asked his followers to send in money to help support his broadcasts. He told his listeners that we were building a radio ship to broadcast the word of God throughout the world and do missionary work. And he asked them to donate money to help pay for the costs of the project. So from the beginning, I knew that the funds would probably come in a little bit at a time.

Brother Stair and Scott hired me for the project because I am very good at building radio stations on small budgets. I'm a real scrounger who can buy transmitters, ships, and equipment cheaper than anyone else.

When you outfit a ship, it's like outfitting a small island community that has to exist on its own in the ocean. You need everything from sponges to bars of soap, food, fresh water, and spare parts for the transmitter. So there was always something we needed to buy. But Brother Stair was good about it. When I called and said that we needed money for something, he usually came up with it. So we got by, and the radio ship gradually came together.

Scott was still in Nevis/St. Kitts, and we'd had no word from him. He was 90% sure that they would agree to allow the first privately owned, 100%-legal-without-question offshore station to broadcast from their waters.

We had to get the major systems operational so we could take the *Fury* down to South Carolina for its final refitting and paint job. Peter and I went down to the engine room and spent a day getting fuel into the big 12-cylinder, 1,000-horsepower diesel engine. Then when we tried to start it, we discovered that the *Fury* had a hell of a lot of broken pipes and castings. Apparently, the people who owned the ship before us did not bother to winterize it and drain the fresh water out of the pipes. We finally got everything fixed, but it took a lot of time and cost about $7,500 to $8,000 that we didn't expect to put out for the parts and labor.

At the end of August, we received bad news from Nevis/St. Kitts. It was no go with that country. After Scott spent six weeks trying to negotiate a deal, they turned us down — and they wouldn't even tell us why. So we had to look elsewhere. Maybe Antigua, Belize, or even Cuba.

It seemed that we had an enemy who was sending out bad reports about us. I think someone in the United States went on a crusade to discredit the operation. Scott did some investigating, and he found out that someone was constantly calling and sending faxes to Nevis/St. Kitts saying that Brother Stair was no good, Scott was no good, I was no good, and claiming that the radio ship project was nothing but an illegal farce.

It really took us by surprise. I felt bad for Scott because he spent so much time and money in Nevis/St. Kitts trying to negotiate the deal. Then some evil bastard saw to it that it didn't go through.

Who it was, we may never know. Brother Stair could have some enemies. He has a lot to say against the New World Order, the Pope, and a lot of the mainline Protestant denominations because he feels that they have all strayed away from the will of God. Or it could have been someone who didn't like me or Scott. But someone just kept badgering Nevis and St. Kitts until it got to the point that they didn't want anything to do with us.

Brother Stair sent us another infusion of cash in early October. So the project went on. The ship now had Belizean registry and a license for our transmitting equipment. It cost nearly $9,000 — but at least we wouldn't have to worry about that problem anymore.

At the end of October, the heat exchanger was in, and Peter and I had worked on the main engine. Brother Stair wanted us to bring the ship down so that the people at his ministry could help with the final preparations. And there was still much to do to get the *Fury* ready to go down to South Carolina for the paint job and final refitting.

The *Fury* set sail at 7:00 AM on November 5 for South Carolina. Frank, Scott, Tom, and the crew were on board. I got sick with the flu and left the boat. By then, I was so ill that I fell off the dock on my way to the car.

I got a call from Scott around 1 PM on November 10. The ship had arrived in South Carolina. All was well, and the main engine worked great.

According to Frank, the ship needed another hundred tons of ballast. He wanted an additional $3,500 to do the job, and was working that out with Brother Stair.

Tom Barna and Brother Stair's people were seasick the whole way down. The Coast Guard boarded them for an inspection on the second or third day, but we had no real problems with them.

I talked to a fellow about doing some engineering on the station. I ran an ad in *Radio World,* and got responses from some people who were willing to go down and spend a month or two working on a radio ship in the Caribbean. I planned to drive down to South Carolina the following week. Hopefully, I'd feel better by then.

Upon arriving in South Carolina, I spent a week on the ship, working on the transmitters from 9 AM to 11 PM. There were sparks, exploding capacitors, blasted feed-through insulators, and a dozen other problems. The smell of burnt resistors filled the air. I tried to fix, bypass, hot-wire, and jury-rig all that I could.

The transmitters we bought had been in storage for about 20 years, so they really needed a lot of cleaning and piecing up. Just getting the filaments to light was difficult. And when I tried to get the 10,000-volt high-power supply to work, one of the capacitors exploded like a grenade and threw shrapnel all over the place. But before the week was over, I got a couple of the transmitters to work pretty well into the dummy loads.

Scott, Tom, and I went to see Brother Stair on his farm. It was a very nice visit. Even though I do not agree with all of Brother Stair's ideas, I believe that he has a true commitment to the Lord.

I left Brother Stair's farm and headed back to New York. As soon as I could hire someone to go on board as the ship's transmitter engineer, they could finish the rest of the de-bugging. Then I could get on with the task of finding people to lease time on our transmitters.

Brother Stair provided a lot of workers to help with the radio ship. In fact, he had half of his ministry chipping paint, cleaning, and moving things around. One day, he had 20 or 30 people from his organization and some other guest churches chipping old paint off the ship.

Walter, one of the fellows from the Overcomer Ministry, was fantastic — he could do anything. He's a master welder, he could build things out of steel, and he knew how to deal with diesel engines and electrical systems. He did a lot to help us, and was a fantastic all-around guy.

The M/V Fury, *Charleston Harbor, November 1993.*

I'd been trying to find an engineer to help work on the *Fury's* transmitters in my absence. In early December, I hired a guy from upstate New York, but he seemed to have disappeared. Hard work always scares people away. Scott was going wild buying high-frequency marine sets and global positioning equipment. I still had the flu, but I was trying to enjoy the fall weather and relax a little bit.

J.P. and I tried to hammer some kind of a deal out with our partners at WHVW. The mortgage was way too high, and we were trying to renegotiate it.

Meanwhile, the ship was painted white. And we had a few more Coast Guard hassles — the FCC was putting pressure on them, it seemed.

I worked at WHVW from 9 to 11 and rebuilt the production studio. Cathy and J.P. were still at it, and I was sick of having to referee the situation.

By late December, the Coast Guard was hassling the ship again. I wanted to hire a guy that the people who made our transmitters recommended. But unfortunately, he wanted a lot of money. Scott broke up a fight among Brother Stair's people and Tom said that he didn't want to work on the ship when they were around. At this rate, the ship wouldn't be ready to go out until February.

Elayne and I spent Christmas in the Catskill Mountains at the Point Lookout Inn and had a pleasant time.

January 1, 1994, ushered in a brand new year, and it started off with a bang. Cathy quit WHVW to work at another station. So now on top of all the other problems, we'd have to find another sales manager.

Scott and Brother Stair locked horns during the first week of January. Brother Stair wanted Scott to leave, and all work stopped on the ship. Another fine mess. I really needed to go down there and get things going again.

I went to the boat on January 8 and tried to work out the problems between Scott and Brother Stair. Scott decided it was best to leave for a while, and work resumed on Monday the 10th. I was busy trying to fix all manner of things.

Then, on January 19th at 8:30 AM, the black day of death came when the bastards from the FCC raided the ship. 48 hours later, she was stripped of all her transmitters, studios, antennas, test equipment, and anything else the government cared to steal.

It seemed that someone in the Charleston area broadcast some tapes of Johnny Lightning's programs the preceding month, and the government claimed that it came from the ship — which was impossible because our shortwave transmitters weren't even hooked up! It really looked like a set-up to me.

Freaked out by the raid, Tom and I spent the day at Brother Stair's farm.

I left for New York City on February 21. The following week, I filed a declaration with the Boston court stating that nothing illegal was done on the ship. The government was trying to get Scott on some kind of conspiracy rap. We needed to find a lawyer soon so we could prove this action on the part of the FCC was illegal.

Brother Stair and the Overcomer Ministries dropped the project fast, but I couldn't really blame them. I guess this was God's way of getting things done. Mixing things up a bit so I wouldn't lose faith.

God forgive me, but sometimes I felt such anger and hatred for all that were responsible for the destruction of our floating radio station. Part of me wanted to say, "May they all burn in hell for this," but I really would not want anyone to suffer. However, what they did was wrong, and they must be stopped from doing it again. I prayed that we would all have the strength in God's will and insight to do the right thing.

I returned to the farm in Maine in mid-February. The snow was at least three feet deep, and almost all powder. I was tired and depressed about all that had happened.

I needed to get back to WHVW and do what I could — but the peace and quiet solitude of the farm were needed, even though it was no fun to be without running water (the pipes were frozen). A warm house and a hot shower are just as important to me as cookies and milk.

Winter lasted way too long. There was lots of snow and exceptional cold. But it ended fast, and spring came early for a change. Scott had been in continual negotiations trying to sell the *Fury,* and thought he found a Haitian buyer.

Johnny Lightning tried to buy some air time on WRNO and Joe Costello turned him down. But WWCR said, "Maybe." In the meantime, I offered him a Saturday night slot from 10 PM to whenever for his "A Little Bit of Everything Show" on WHVW. He started on May 7. So the station went on with no money, little sales, but a lot of heart. Tom was working at WREM, basically idling the station and airing some talk radio programs.

On April 24, I made the front page of the Charleston, South Carolina newspaper, *The Post and Courier*. They did a story about the ship, Brother Stair, Scott, and me. The FCC told the reporter that they didn't prosecute me because they didn't want to turn me into a martyr — the best laugh of the year. The government's lies never end. I sent a packet of stuff to the South Carolina ACLU. Hopefully, they could stop the madness.

All I wanted to do was broadcast messages of peace, love and understanding to the world. Was that such a terrible crime?

Chapter Fourteen
Returning To Maine

In the spring of 1994, things still looked pretty bleak at WHVW. I tried my best to get that station on track, but the whole deal was dragging me down. I just couldn't fight it anymore — it took up far too much of my time and energy. I knew it was high time to get the hell out of New York and move back to my farm in Maine, where it was peaceful and quiet. So late that May, Tom and I loaded all the stuff from my house in Yonkers into a 24-foot Ryder truck and drove up to Monticello.

Around that time, I made another major decision — I decided to try for my shortwave license again. I'd had enough of this government bullcrap. If I was going to spend my time and money fighting the FCC, I wanted to get something in return. So I called up my attorney, Michael Couzens, and told him of my plans. He wanted us to meet face-to-face with the commission. I agreed, and we set the meeting up for late July.

When we got together with the FCC, things went about as well as expected. They confirmed that the taped RNI broadcasts that went out over the air in December and January were what triggered their actions against the ship. I told them that I had nothing to do with those transmissions, and that I had done everything I possibly could to keep things legal and proper with the *Fury* project. (To this day, I have no idea where those broadcasts came from. I only know that we had nothing to do with it.) I think they believed me. But after our discussion was over, I still had no real sense of where I stood with the FCC in regards to them issuing me a license.

In late 1994, Elayne decided that she also wanted to get away from all the problems of living in the New York area. So when I had some time off, we drove around and looked at property in the southern part of Maine. Finally, we found a nice house in Kennebunk. It was a fine home with a big garage — a bit expensive, but worth it. Elayne was ecstatic. Now she would have a nice house to live in — and I would have a base of operations when I had engineering work in that part of the state. It also turned out to be a good location for selling some of my antique electronic gear.

January 1 brought a new year full of hopes and aspirations. Certainly this year would be better than the last. How could it not be? I hoped and prayed for peace and love in the year of our Lord 1995. Hopefully, Elayne and I would be settled in Maine, and all radio matters would work out OK.

Yes, a new year was here — and I was thinking about broadcasting from the sea again. Tom and I went to Camden, New Jersey, to look over an old lightship which I wanted to convert to an offshore broadcast station. I know it sounds crazy, but somehow I feel naked without a ship. The ship needed just about everything, but at least the hull was in good shape. So I made an offer and waited to see what would happen.

At the time, I was trying to find a way out of the Hyde Park situation. The specter of doom over losing my investment did not make me happy. I put the station up for sale with a broker, but with WHVW's financial situation being so dismal, I wasn't optimistic about finding any takers. But I had to try something.

WREM was also causing me a lot of problems that winter. In the middle of January, the transmitter went out. I trudged out to the building in sub-zero weather to discover that it had a blown modulation transformer — absolutely the worst part to blow out in the worst part of the year. Fixing a 5,000-watt transmitter with a $0 budget wasn't easy. I had to scrounge around all over the place to find the parts I needed. But that's my specialty — fixing, designing and building radio stations with little or no money. Then, to top it all off, the guy who was leasing the station from Dr. Rish skipped town, sticking us for about $6,000 worth of unpaid bills.

Aside from all that, I had some very interesting dreams that winter. Throughout my life, I've always been a prolific dreamer. And these dreams were especially nice because they took me to a place where everything made sense and all was wonderful. The peace, love, and beauty were so intense that I knew they must truly be gifts of higher levels of awareness from God. Perhaps it's just that I've trained myself to remember my dreams more than the average person. But I sometimes have really gorgeous dreams — and I wake up feeling wonderful.

In the spring of 1995, it appeared that we were going to have to move WHVW. My father, who was representing J.P. in that action, tried to get the Poughkeepsie courts to grant us a stay. On March 9, the case went to court. The judge gave the owners their property back, but not the radio station, license, or any monetary judgment. Dad thought there was a chance we could work out a lease arrangement with them, but they turned us down.

As the deadline for the station's eviction approached, Dad, J.P., and I tried for an 11th-hour bailout by making an offer on the property, but they didn't go for it. WHVW was officially evicted on April 15, almost three years to the day from the time that we purchased it. And we had to move the station from 507 Violet Ave. to an office building in downtown Poughkeepsie.

I worked my buns off erecting a 60-foot tower on top of the building. I used a self-support tower and raised it up in one piece. Then I jacked the whole assembly up and put a base insulator underneath it. As expected, the signal levels were way down. But with only seven radial wires on the roof, it's amazing the signal got out at all.

Around that time, Johnny Lightning and I decided to talk to WWCR and see if they would lease us a Sunday-night time slot so we could reactivate Radio New York International. The only thing they had available was from midnight to 1 AM. It was pretty late — but we took it. The first program went out over the air on May 1. Tom and I talked about the *Fury* project and told everyone how the government came down to destroy the station and steal all of our equipment. It really felt good to have RNI back on shortwave again.

Of course, I still wanted RNI to have its own transmitters. And with the FCC being so stubborn about not giving me a license, offshore radio seemed to be the only way to go. So, on May 11, I went to New Jersey to sign the papers and buy the lightship.

Now that I had a ship to outfit, I sat down and thought about how to promote it. First of all, I decided to call the venture Lightwave Mission Broadcasting. I placed a few ads in *Radio and Records, Broadcasting*, and *Offshore Echo's* magazines to see if anyone was interested in investing. And my friend Anita McCormick posted some announcements about the project on the Internet. I planned to attend the *Offshore Echo's* Free Radio Conference in France and see if I could find someone who might be interested in investing in the project, but unfortunately I wasn't able to go — I didn't have the extra time or money.

When I went back to Hyde Park, J.P. had some good news for a change. He told me that he and another guy had decided to buy into WHVW. They promised to pay us some on it every week. So at long last, Dad and I had hopes of getting at least part of our money back out of that crazy fiasco.

In between all of this, Johnny Lightning and I did our midnight RNI shows on WWCR every Sunday evening. But after about two months, we decided to cancel. It was a terrible time slot. We did get a lot of calls and quite a bit of mail from our listeners. But the expenses were running about $100 a week, and we weren't getting anything near that in donations. Besides, we were tired of paying someone else to air our shows when it ought to be the other way around. George McClintock said that the Woodstock mentality doesn't work anymore. Talking about love, peace and understanding just drives some people crazy, I guess.

In July, a guy named Ralph and some people he knew in Houlton approached me about leasing WREM and turning it into a rock'n'roll station. Ralph paid off most of the past-due bills and Rock 710 made its debut the very next day. Making an AM station into a rock'n'roll voice was quite a bold move. It was really refreshing to see such enthusiasm in radio.

That summer, I sent out some flyers advertising my services as an engineer and got a gig at WSME and WCDQ in Sanford, Maine. Finally, I had some paying work. And after months of searching, Elayne got a job working at an antique gallery in Wells, Maine. So the financial picture was finally starting to look better for both of us.

I tried to promote Lightwave Mission Broadcasting all summer. But when I hadn't received any serious responses by September, I pretty much gave up on finding an investor for the radio ship project and decided to sell the vessel. Elayne and I visited the Shore Village Museum in Rockland, which has one of the world's largest displays of lighthouse lamps, lenses, and other apparatus. I spoke to the curator about selling the lightship. He gave me the names of some people who might be interested in buying her. She was a nice ship, and I wanted her to have a good home.

Another one of my goals that fall was to help Elayne get into the radio business. We looked at stations all over Maine, but everything we saw was too expensive. Then WEGP in Presque Isle came up for sale. It needed a lot of work on the transmitters, tower, building, etc. But we talked to the owners and worked out a price and a payoff schedule which enabled us to buy the station over a five-year period. It was affordable, so we took it. I'm at my best when I'm building something, so I knew that this project would be good for both of us.

Elayne filled out the papers to purchase WEGP in December. At the time, we were still deciding whether we should lease it out or run it ourselves and do a talk-radio format. Eventually, we decided to run it ourselves. Talk radio was sorely needed in Aroostook County — and WEGP was the perfect station for it. It was a challenge to build a radio station with the tight money situation we were under. But we did have the transmitter site, a house in which to build the studio and office, and plenty of equipment.

WEGP was the first commercial station that hired me to do engineering work after I got out of college. And now I had a chance to engineer it again and bring it back into the Presque Isle market with Elayne as the owner and manager of the station. I guess as things go around, so do they come around.

A few days later, I went to visit the lightship for the last time. My heart ached to stand on her deck and think of all that she could have become. But I had a station to engineer, so I directed my thoughts back to that project.

WEGP was not just fun and games. It had to make some money, or else we'd be cooked. I was temporarily out of engineering work, WHVW was not paying what they owed me, and the ship wasn't selling. So things were tough. I built, I prayed, I hoped, and I dreamed.

One of the first things I did when I got back up to Presque Isle, Maine, was to look over the house at 3 State Street Place where we planned to build WEGP. Despite some broken water pipes and other assorted problems, it looked like a good quiet location for a radio station. There was just enough room to build a broadcast studio and put in an office or two.

Moving equipment into WEGP's State Street location was my life for a few days in mid-January. Gary Stone and I moved the 17-year-old WTMS automation system to WEGP. Strange, because I saw it roll in brand new into the WTMS studio in Presque Isle back in 1979. The temperature was below zero, and the frigid weather sure slowed things up.

In late January, I got a call from a chap in Holland. He said that some group in Israel was outfitting an old British lightship as an offshore station and they needed an engineer. But nothing ever came of that. Around that time, I also got a call from a fellow who said that he was interested in buying the old lightship.

By then, I had no money to do anything but live. So there I sat in a broken-down trailer in beautiful northern Maine, fixing up a broken-down AM radio station in hopes of not going broke. Working alone that January, I built a studio at State Street Place in two days at a total cost of $100. It looked like an antique that belonged in the Smithsonian Institution — but it worked!

I went out to the old, defunct WKDX transmitter site in Presque Isle, Maine, to salvage some parts. Then I drove to WHVW in Hyde Park to pick up some equipment and the old 10-foot Mutual satellite dish. As expected, the station was a mess and still wasn't making any money. So using whatever equipment I could scrounge up, I moved ahead with the engineering of WEGP. Elayne was very supportive and helped me out all the time.

On Valentine's Day, Ralph threw in the towel with WREM. He owed a lot on the electric bill, and said he wouldn't pay it. So Dr. Rish, my father, and I decided to take the station back and run it with a rock'n'roll format to see what happened. Larry would do sales, Jay would do programming, and Linda would handle operations. We seem to be on some sort of a quest with WREM to make just one dollar out of the place.

Late that February, I finally found a buyer for the lightship. And a great mass of iron was lifted from over my head. Perhaps it was foolish of me to assume that someone would just come forward and finance my project. But with the FCC dragging their feet on my license application for who knows how long, I just HAD to try something.

By mid-March, I'd finished WEGP's main studio and everything seemed to be working well. But we still needed a 12-foot satellite dish to pick up the various talk networks. I finally found a satellite dish we could use and put it up — and for three days, I tried in vain to get it to work correctly. The C5 satellite is very low in the sky, and I thought the problem might be that we had too many trees in the way. But the satellite receiver turned out to be a dud, and I had to send it out to Colorado for repair.

We had originally planned to have WEGP on the air by mid-April. But the combination of bad weather and financial problems put us about a month behind schedule.

On May 15, WEGP finally went on the air. Despite a glitch here and there, the equipment worked out fine. And I did my first live talk show in a long time. We hired a woman to help us with sales. She was young and inexperienced, but willing to give it a try. My sister Barbara also offered to do some selling and help out with the station.

Elayne and I only had to put out about $5,000 to $8,000 to get WEGP on the air. It is a truly remarkable thing to put a radio station on the air with such a low budget and hodgepodge of equipment. As of mid-May, 1996, Elayne and I are running WEGP. So far, the station is working out very well. WEGP and its talk-radio format has been enthusiastically received by the Presque Isle community. New advertisers are signing on every single week. And I really enjoy having the opportunity to do a morning talk show again.

So even after all that had happened with the FCC, I'm still doing radio. And I am doing it on my terms. Granted, it is hard work. When you run a radio station with only one other person (Elayne), it basically means working 14-hour days. There are no holidays. When you're in the radio business, it's a 7-day-a-week affair, because radio stations never sleep.

But despite all the problems, it was wonderful fun to resurrect a radio station, and I would encourage anyone out there who wants to run their own ma-and-pa station to do it. Local radio is where it's at. It's just like running a corner grocery store. You get to know people and be an important asset to your community. And you can get into the electronics end of it just as deeply as you want.

There are over 10,000 AM and FM radio stations in this country. And chances are you can find one that is dark or in financial problems for a reasonable price. Using some ingenuity and the equipment at hand, you can get on the air and have fun with what I have found to be the most fascinating of all businesses — radio broadcasting.

As I write this, I'm anxiously waiting to hear from the FCC about my shortwave license application. My government, it seems, is treating me like a second-class citizen. But hopefully, they will change their attitude and make a favorable decision to allow me to build an international broadcast station on my farm in Monticello, Maine — as I have wanted to do for so many years now.

On my shortwave station, I would air music, news, and information to a world-wide audience. I would lease time out to groups or organizations at a reasonable cost, so that people who normally couldn't afford to buy radio time would have a chance to express themselves. And I'd like to give all the people that have been involved with Radio New York International, past and present, a chance to return to the airwaves.

In short, my station would be a free and uncensored voice — something that is sorely needed at this time when there are so many angry, whining, hateful voices on talk radio. The world needs an RNI. A moderating voice, a friendly voice, maybe a nice voice. A voice that says, "Hey — I'm okay, you're okay, and we're okay."

To this day, the most wonderful communications medium, the medium that touches our hearts and our souls like no other and paints glorious wonderful pictures in our minds, is still radio.

And as I have said many times before, I will gladly engineer and build a radio ship for anyone who can finance the project for any good and peaceful purpose. And I will build it cheaper, better and faster than anyone else. Having one completely free and uncensored radio ship out there legally broadcasting from someplace on earth will always be a goal, a dream, and a hope of mine.

— Allan H. Weiner

Appendix
Access To The Airwaves
In The 21st Century

When I was a teenager back in the 1970s, pirate radio was the only way to get access to the very commercially dominated airwaves. And I guess you could still make the same argument today. But things have changed somewhat. People do have some alternative ways to go on the air without running into trouble with the FCC.

One idea is go to the operator of a small local radio station and say, "Listen, I'm interested in doing radio. Do you have any time slots at night when you would let me on the air?" You have two options here. If the station is looking for programming, they might just let you DJ or run the show for them for nothing. If you're lucky, you might even get paid for it. The other option is buying time. A lot of small to mid-size AM stations have air time that they would consider selling at rates between $25 to $100 an hour, depending on the time, power of that station, geographical location, and so on.

Another possibility is to secure some time from a cable company. A lot of cable systems have a channel where community announcements and swap boards scroll on screen all day. They usually play music in the background, and they might be willing to lease that channel out for you to do a program. They might also be willing to lease one of their FM carrier channels out to you. It's legal, and it's a good way to get on the air on a tight budget, and get your message across without worrying about a government raid.

You might also want to check out the possibility of obtaining a local management agreement — LMA, as they call it now. People are doing this constantly. And I think it's a good alternative. You can rent out the use of a station to do your own programming. You can sell ad time or lease out part of the broadcast schedule to other organizations to help pay the bills.

And you might want to get together with a few other people and look around for a small-town AM or FM radio station to buy. There are over 10,000 commercial stations in the United States, and a good number of them aren't doing very well financially — especially AM stations in small-town markets. They might be bankrupt. They might even be off the air. You just walk in, talk to the owner, and see if you can work a deal. With a little bit of financial prowess and good bargaining, you might be able to purchase the station for the cost of the bad debts or the cost of the property.

So if you are really into doing something like this, purchasing a broken-down radio station just might be the thing that gets you on the air — whether you want to play jazz, country, alternative rock, or have a talk format. More and more people are doing it. And I think it's a good thing. There will always be opportunities for that kind of access with the thousands and thousands of radio stations out there.

I know that it's easy to get hold of a ham transmitter, put a station on, and broadcast to a few shortwave listeners scattered throughout the country. But with the illegal search and seizure tactics that the FCC uses these days, you can seriously jeopardize yourself and your equipment by doing so. If they track you down, the FCC will come into your house with federal marshalls armed with guns to take away your equipment. Even worse are the fines of $10,000 or more that the FCC can impose on anyone they catch broadcasting without a license. And the fines don't go away. If they can prove the charges, you will be liable for it.

YOU WILL ALSO WANT TO READ:

☐ **70050 PIRATE RADIO OPERATIONS,** *by Andrew Yoder and Earl T. Gray.* Pirate radio is one of the Communications Age's most fascinating developments! Now, for those hobbyists who yearn to learn the ins and outs of clandestine radio broadcasting, there's a wealth of knowledge available in *Pirate Radio Operations!* For the first time, there's a hands-on manual that fully explains the intricacies of this burgeoning pastime. Yoder has devoted his energies to pirate radio for years, and now he shares his practical expertise with the world. Complete with numerous photographs and illustrations that provide workable designs and schematics for all pirate radio buffs, this is the finest how-to book ever published on this subject. *1997, 5½ x 8½, 376 pp, illustrated, soft cover.* **$19.95.**

☐ **10052 CODE MAKING AND CODE BREAKING,** *by Jack Luger.* We live in an information age; information is bought, sold and stolen like any other good. Businesses and individuals are learning to keep their secrets safe with this practical, illustrated guide to building and busting codes. Learn how to construct simple or complex codes. Learn how computers are used to make and break codes. Learn why the most unbreakable code isn't always the best. Ideal for those interested in professional and personal privacy. *1990, 5½ x 8½, 125 pp, illustrated, soft cover.* **$12.95.**

☐ **91085 SECRETS OF A SUPER HACKER,** *by The Knightmare, Introduction by Gareth Branwyn.* The most amazing book on computer hacking ever written! Step-by-step illustrated details on the techniques used by hackers to get at your data including: Guessing passwords; Stealing passwords; Passwords lists; Social engineering; Fake e-mail; Viruses; And much more! The how-to text is highlighted with bare-knuckle tales of the Knightmare's hacks. No person concerned with computer security should miss this amazing manual of mayhem. *1994, 8½ x 11, 205 pp, illustrated, soft cover.* **$19.95.**

Check out our catalog ad on the next page,

Loompanics Unlimited
PO Box 1197
Port Townsend, WA 98368

AAW7

Please send me the books I have checked above. I have enclosed $_____ which includes $4.95 for shipping and handling of the first $20.00 ordered. Add an additional $1 shipping for each additional $20 ordered. Washington residents include 7.9% sales tax.

Name _____

Address _____

City/State/Zip _____